A Dictionary of Scientific Quotations

The harvest of a quiet eye,
That broods and sleeps on his own heart

From *A Poet's Epitaph*
by William Wordsworth

A Dictionary of
Scientific Quotations

by Alan L Mackay

With a Foreword by the late Sir Peter Medawar

Institute of Physics Publishing
Bristol and Philadelphia

IOP Publishing Ltd has attempted to trace the copyright holder of all the quotations reproduced in this publication and apologizes to copyright holders if permission to publish in this form has not been obtained.

British Library Cataloguing in Publication Data

A dictionary of scientific quotations.
I.Mackay, A.L.
503

ISBN 0-7503-0106-6

Library of Congress Cataloging-in-Publication Data are available

A Dictionary of Scientific Quotations is a second and expanded edition of *A Harvest of a Quiet Eye* which was first published in 1977 under the Insitute of Physics Imprint.

Reprinted 1992

Published by IOP Publishing Ltd, a company wholly owned by the Insitute of Physics, London

IOP Publishing Ltd
Techno House, Redcliffe Way, Bristol BS1 6NX, UK
US Editorial Office: IOP Publishing Inc., The Public Ledger Building, Suite 1035, Independence Square, Philadelphia, PA 19106

Printed in Great Britain by Galliard (Printers) Ltd, Norfolk

Foreword to the First Edition

I was charmed and delighted by this collection of aphorisms and quotations, and hope and expect that many others will be too. Most of them will appeal to scholars generally—very few are for scientists alone, though these few are well chosen (it is fun to read Le Chatelier's theorem as it was propounded by the master himself—(L:35)).

As is usual with compilations of this kind, some quotations or aphorisms are so persuasive and so well put that one wishes one had said them oneself; others are wrong-headed—even Hardy nods (H:29)—and others still tell us more about their authors than about their subjects: Hilaire Belloc (B:50) cannot have intended to make a public exhibition of himself, but in one short passage he skilfully betrays a total incomprehension of the scientific process, based upon that archaic usage of the word 'experiment', according to which an experiment is an answer to the question: 'I wonder what would happen if?' and again (D:5) the showbiz or cocktail party air of Salvador Dali's comment on DNA makes other quotations seem by comparison more profound than they really are.

The inclusion of a few graffiti was a stroke of genius; my own favourite wall decoration is to be found in the Faculty Club of Rockefeller University, a place where reputations are keenly debated and appraised: a cartoon shows three or four eager scientists discussing the claim to fame of, one presumes, Prometheus: 'Sure, he discovered fire', the caption runs, 'but what has he done *since*?'

At first sight some of the entries strike the reader as irrelevant to science or scientists, but on closer inspection they will be found either to have a sting or to accord with a train of thought relevant to Dr Mackay's own personal selection of extracts—the labour of many years and clearly a labour of love.

I propose to add one more epigram to this ostensibly irrelevant category: in these days of cost-consciousness, when the funding of research is being administered as if scientific research were a branch of the retail trade, and when 'pure science' and those who practise it are coming under an increasingly cynical scrutiny, it is as well to remember the definition that Oscar Wilde puts into the mouth of Lord Darlington in *Lady Windermere's Fan* : a cynic is 'a man who knows the price of everything and the value of nothing'.

It is strange in any book of quotations not come upon a great spout of steamy spray from the man Logan Pearsall Smith described as the great Leviathan of English letters, but this is because Dr Johnson did not often bask offshore of the New Atlantis, though when he did so, his observations had a strength, sanity and gravity that will instantly recommend them to the increasing number of younger scientists who are deeply concerned to introduce a moral valuation into science and its applications. In his *Life of Milton* , Johnson chides Milton (and incidentally Cowley) for thinking of an academy in which the scholars should learn astronomy, physics and chemistry in addition to the common run of school subjects. Johnson did not approve of these schemes, for:

> '...the truth is that the knowledge of external nature and of the sciences which that knowledge requires or includes, is not the great or the frequent business of the human mind. Whether we provide for action or conversation, whether we wish to be useful or pleasing, the first requisite is the religious and moral knowledge of right and wrong; the next is an acquaintance with the history of mankind, and with those examples which may be said to embody truth, and prove by events the reasonableness of opinions. Prudence and justice are virtues, and excellencies, of all times and of all places; we are perpetually moralists, but we are geometricians only by chance. Our intercourse with intellectual nature is necessary; our speculations upon matter are voluntary, and at leisure. Physical knowledge is of such rare emergence, that one man may know another half his life without being able to estimate his skill in hydrostatics or astronomy; but his moral and prudential character immediately appears.'

Sir Peter Medawar
Clinical Research Centre, Harrow

Preface to the First Edition

Scientists have often been reproached for their apparent unfamiliarity with the rest of our cultural inheritance. Inasmuch as science represents one way of dealing with the world, it does tend to separate its practitioners from the rest. Being a scientist resembles membership of a religious order and a scientist usually finds that he has more in common with a colleague on the other side of the world than with his next-door neighbour. But as Shakespeare said of the two cultures: 'Hath, not a Jew eyes? Hath not a Jew hands, organs, dimensions, senses, affections, passions, fed with the same food, hurt with the same weapons, subject to the same diseases, healed by the same means, warmed and cooled by the same winter and summer, as a Christian is?' In this case the division was by religion rather than by attitude to science.

Scientists do live in the real world and share in developing and unifying its culture. Many scientists, especially biologists, are able to communicate more widely than to their professional colleagues their sense of wonder at the workings of nature and indeed, for a creative person, the form of the scientific paper is so constrictive that another outlet for his writing is necessary. This collection of quotations is intended to show the wholeness of our culture by demonstrating that scientists contribute to the humanities, and that from Chaucer to Auden, the great humanists have also been concerned with science in all its aspects.

A quotation is a polished prefabricated unit of thought or discourse which has many connotations and associations built in to it. It is thus like the text for a sermon, serving as a point of departure for many lines of thought. Each of us knows many thousands of words and can give, for almost any word, a definition close to that to be found in a dictionary. Yet each one of us has only ever looked up perhaps ten per cent of all the words he knows. We have learnt words by picking them up in their contexts. Each transaction with a word polished it and defined its use and meaning more exactly. Words are coupled into phrases which carry complete thoughts, associations and meanings. We have the subjective feeling, which probably reflects a genuine physical basis, that words are stored in our brains in a vastly ramified network. Extraction of a particular word stirs others and whole phrases and sentences follow. Quotations are, in effect, thoughts embedded in memorable phraseology and polished by use. They are large preformed elements and necessarily combine deep structure (the ideas) with surface structure (the actual words in which the ideas are caught). It is easy to incorporate them, like plug-in circuit boards, into one's own thinking machine.

This is a work of pure plagiarism and I have gleaned items from wherever I came upon them, finding texts sometimes corrupt and full references usually lacking. I apologise for inadequacies and hope that enthusiasts will correct any mistakes and will be stimulated to contribute further entries for later editions. Not all entries in this selection will be familiar to the average scientist, but as quotations are for use, it may help those who write and speak about science to illustrate their material with them and thus some of the less familiar may take root and propagate themselves. Different people will be led along different pathways of thought and some may be stimulated to seek further acquaintance with authors new to them.

In a way the compilation of this book has long been inevitable, because, in 1940, my classics master, S G Squires, required that each pupil in his class should keep a notebook for quotations—I still have mine. So each morning, while the British Empire crumbled, we learnt a new Latin tag and were tested on them once a week. I had doubts of the value of the Classics and even of Shakespeare, but the influence at that formative age had its effect. At the same time the endless exposure to the Bible and the Liturgy of the Church of England provided the essential basis for an informed rationalism and a feeling for the cadences which underlie most of English prose. From the same period, H C Palmer, my senior science master, convinced me that science was interesting and important and set me on a scientific career.

Later, as a prize for the first year examination in Natural Sciences in Cambridge, which was unprecedentedly postponed because it had been fixed for what turned out to be VE-day, I chose *The Social Function of Science* by J D Bernal. This was my introduction to the works of the marvellous group of encyclopaedists which included Bernal, Waddington, Needham, Haldane, and many other less well-known figures. This book was a revelation as to how areas of life which had been disconnected actually fitted together. Later still, after work in industry, I was able to join Bernal's crystallographic laboratory at Birkbeck College, London.

The years in which Bernal was active were immensely stimulating. Besides the revolution in biology all kinds of social, scientific and political movements had their base there and the decrepit buildings which housed the laboratory were an important international and intercultural crossroads. Alas, in 1963, Bernal suffered a stroke and he passed his last years tragically cut off by his own failing senses. Bernal's example of the excitement and wholeness of life remains an inspiration.

However, it is to my father, whom a book of poetry helped to carry through the Great War, from September 1916 to the end†, that I would wish to dedicate this selection. He lived to see a great-grandchild. (1976)

Alan L Mackay
Department of Crystallography, Birkbeck College, London

---

† In his war diary for 22 August 1917, at the Battle of Ypres, he wrote, 'Zero Hour was 4.45 a.m. and was a sight I will never forget. God knows how anybody got over at all.'

Preface to the Second Edition

Quotation and citation is intrinsic to science. The first number of *Nature* (4 November 1869) included a set of aphorisms selected by T H Huxley from Goethe†.

Having collected as far as possible the familiar sayings which everyone concerned with science ought to know, I have widened my scope and have tried to include some new items which may be useful to people who write and speak about science and society. Sampling what is actually quoted shows that there is only a finite population of quotations in circulation. In searching, the same items come up again and again. Here and there I have again indulged my own taste for the irreverent and paradoxical.

Perhaps some of these items can be used as pinpricks to Philistines of the Thatcher regime which has so blighted the last decade. The 1980s ended with revolutions which brought more hope for the intrinsic humanity of humankind than for many a long year. These hopes have largely evaporated in subsequent chaos. We see the rise of nationalisms and religions, largely incompatible with the internationalism of science and with the increasing interdependence of world societies and economies. In Britain we have perhaps escaped fascism but, with the 1990s, the world enters a period of chaos in which it is more necessary than ever before to think clearly about what we are doing and where we want to go.

Since 1990, the world situation has deteriorated. The Gulf War has taken place. People have been slaughtered by modern science and technology as at the Battle of Omdurman (1898) (see the entry under Churchill) or the Battle of Gök Tepe (1881). The casualties in Iraq compare with those which occurred in 1258 when Baghdad was sacked by the Mongol armies of Hulagu and a pyramid of 100,000 heads was made. The irrigation system was then also destroyed as the oil fields of Kuwait this year. The trade in armaments will continue to grow with the advertisement afforded. Half our scientific effort will continue to be devoted to arms for export and for defence. The population crisis, forseen, as much else, by H G Wells, still looms, to be complicated by the spread of AIDS.

Scientific analysis is more than ever necessary. Mathematical logic might be thought by monetarists to be one of the least important departments of a university. Yet the Polish school of this abstruse philosophy won the war for us by solving the Enigma machine. Similarly, the most abstract mathematics is now leading to an understanding of chaotic phenomena in economic and other systems.

---

† See G Beer 1990 *Notes Rec. R. Soc. Lond.* **44** 81–99.

In Britain, it has been evident to all, since the Great Exhibition of 1851, that the prosperity of these islands depends on industry and agriculture, that these stand on technology which stands on science and that science is based on the general education of the population.

We hope that this dictionary will, in a humble way, help with the inculcation of scientific attitudes into our civilisation and the unification of the two cultures.

Alan L Mackay
Department of Crystallography, Birkbeck College, London
1991

Phillip Hauge Abelson 1913–

1   Part of the strength of science is that it has tended to attract individuals who love knowledge and the creation of it. Just as important to the integrity of science have been the unwritten rules of the game. These provide recognition and approbation for work which is imaginative and accurate, and apathy or criticism for the trivial or inaccurate.... Thus, it is the communication process which is at the core of the vitality and integrity of science.... The system of rewards and punishments tends to make honest, vigorous, conscientious, hardworking scholars out of people who have human tendencies of slothfulness and no more rectitude than the law requires.
     The Roots of Scientific Integrity Editorial in *Science* 1963 **139** 3561

Russell Lincoln Ackoff 1919–

2   Common sense ... has the very curious property of being more correct retrospectively than prospectively. It seems to me that one of the principal criteria to be applied to successful science is that its results are almost always obvious retrospectively ; unfortunately, they seldom are prospectively. Common sense provides a kind of ultimate validation after science has completed its work; it seldom anticipates what science is going to discover.
     Decision-making in National Science Policy 1968 (London: Churchill) p 96

3   Nature is not organised in the way universities are.
     Probably magazine section of the *New York Sunday Times*

Douglas Adams 1952–

4   [The great supercomputer, asked what is the answer to] the great problem of life, the universe and everything [replied, after many years of computation] 42.
     The Hitch-hiker's Guide to the Galaxy, but see also D Ruelle *The Claude Bernard Lecture, 1989* 1990 *Proc. R. Soc.* **427** 241–8

Henry Brooks Adams 1838–1918

5   The assumption of unity which was the mark of human thought in the middle-ages has yielded very slowly to the proofs of complexity. The stupor of science before radium is a proof of it. Yet it is quite sure, according to my score of ratios and curves, that, at the accelerated rate of progression shown since 1600, it will not need another century or half century to tip thought upside down. Law, in that case, would disappear as theory or *a priori* principle, and give place to force. Morality would become police. Explosives would reach cosmic violence. Disintegration would overcome integration.
     In Wm Jordy *Henry Adams—Scientific Historian* 1952 (New Haven, CT: Yale University Press) p xi

6   The future of Thought, and therefore of History, lies in the hands of the physicists, and ... the future historian must seek his education in the world of mathematical physics. A new generation must be brought up to think by new methods, and if our historical departments in the Universities cannot enter this next phase, the physical departments will have to assume this task alone.
     The Degradation of the Democratic Dogma 1919 (New York: Macmillan)

7   Some day science may have the existence of mankind in its power and the human race commit suicide by blowing up the race.
Letter 1862

8   [After viewing the Palace of Electricity at the 1900 Trocadero Exposition in Paris] All the steam in the world could not, like the Virgin, build Chartres.
The Dynamo and the Virgin in *The Education of Henry Adams* 1918 (New York: Heritage Press) p 434

Homer Adkins

9   Basic research is like shooting an arrow into the air and, where it lands, painting a target.
Nature 1984 **312** 212

Aeschylus fl 463 BC

10   *Prometheus:* My mother ...
foretold me, that not brute strength
Not violence, but cunning must give victory
To the rulers of the future.
Prometheus Bound

11   *Prometheus:* I caused mortals to cease forseeing doom.
Chorus: What cure did you provide them with against that sickness?
Prometheus: I placed in them blind hopes ... I also
gave them fire.
Prometheus Bound

Louis Agassiz 1807–1873

12   Every great scientific truth goes through three stages. First, people say it conflicts with the Bible. Next they say it had been discovered before. Lastly they say they always believed it.

George Agostinho da Silva

13   Science is the meeting place of two kinds of poetry: the poetry of thought and the poetry of action.
1978 in June Goodfield *An Imagined World* 1981 (London: Hutchinson) p 142

Ahmose the Scribe ca 1650 BC

14   In each of 7 houses there are 7 cats; each cat kills 7 mice; each mouse would have eaten 7 hekat of grain. How much grain is saved by the cats ?
[Presumably the origin of the *Mother Goose* rhyme: 'As I was going to St Ives...' in *The World of Mathematics* ed J R Newman, 1956, vol I (New York: Simon & Schuster) p 177

15   Ways of investigating Nature and knowing all that exists, every mystery ... every secret.
[Title of the Rhind Papyrus (on Egyptian mathematics)]
In *The World of Mathematics* ed J R Newman, 1956, vol I (New York: Simon & Schuster) p 177

Mark Akenside 1721–1770

16   Give me to learn each secret cause;
Let number's, figure's motion's laws

Revealed before me stand;
These to great Nature's scene apply,
And round the Globe, and through the sky,
Disclose her working hand.
Hymn to Science in *Works of the English Poets* ed S Johnson, 1779, vol 55 (London)
p 357

Alain [Emile Chartier] 1868–1951

17   There are only two kinds of scholars; those who love ideas and those who
hate them.

Albertus Magnus 13th century

18   Do there exist many worlds, or is there but a single world? This is one of
the most noble and exalted questions in the study of Nature.

Alfonso X (The Wise) [King of Castile and Leon] 1252–1284

19   [On having the Ptolemaic system of astronomy explained to him] If the Lord
Almighty had consulted me before embarking upon Creation, I should have
recommended something simpler.
[Attributed]

Al-Khorezmi 780–850

20   With my two algorithms, one can solve all problems—without error, if God
will!
*Algebra. Compact Introduction to Calculation by Rules of Completion and Reduc-
tion. Kitąb Al-muntasar fi hisab al gabr w'almuqabulun* in *Science Focus (IBM)* no 1
September 1981
[The written numerals are still called algorismos in Portuguese]

Luis Walter Alvarez 1911–

21   There is no democracy in physics. We can't say that some second-rate guy
has as much right to opinion as Fermi.
In D S Greenberg *The Politics of Pure Science* 1967 (New York: New American Library) p 43 © D S Greenberg, 1967

American Philosophical Society

22   ... It shall and may be lawful for the said Society by their proper officers,
at all times, whether in peace or war, to correspond with learned Societies,
as well as individual learned men, of any nation or country ...
[In its charter of 1780]

Andre Marie Ampère 1775–1836

23   The future science of government should be called 'la cybernétique'.
1843

Poul Anderson 1926–

24   I have yet to see any problem, however complicated, which, when you looked
at it in the right way, did not become still more complicated.
New Scientist 25 September 1969 p 638

Edward Neville da Costa Andrade 1887–1971

25   [Of Sir William Henry Bragg 1862–1942] There were many things whose
existence he preferred not to acknowledge. ... There was, we like to think,

something peculiarly British about Bragg. His attitude to physics was that
characteristic of the great experimenters of our land, especially his strong
pictorial sense.
Obituary Notices of Fellows of the Royal Society November 1943, p 4

Yuri Vladimirovich Andropov 1914–1984

26   ... how can we explain that in the 60th year of Soviet power in the USSR
there still exist so-called *inakomyslyashchie* [those who think otherwise].
Pravda 10 September 1977

27   [Warned officials against] the magic power of quotations.
Article in *Kommunist* on the centenary of Marx's death. In *The Guardian* 12 April
1983, p 21

Anon.

28   ... like the statistician who was drowned in a lake of average depth six
inches.

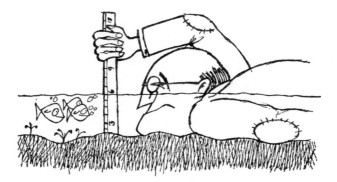

29   'Life is very strange' said Jeremy. 'Compared with what?' replied the spider.
In N Moss *Men Who Play God* 1970 (Harmondsworth: Penguin) p 256

30   'Tis further from London to Highgate than from Highgate to London.
[An example of a non-commutative metric. Highgate is at the top of a hill.]
James Howell *Proverbs* 1659

31   All sugar, cream and glyceryl monostearate.
[On Chaikovski]
The Guardian 1977

32   *amrtam ayur hiranyam.*
Gold is immortality.
Satapatha Brahmana III, 8, 2, 27

33   Being before the time, the astronomers are to be killed without respite; and
being behind the time, they are to be slain without reprieve.
Shu Ching (before 250 BC) in *Nature* **225** 894

34   A biophysicist talks physics to the biologists and biology to the physicists,
but when he meets another biophysicist, they just discuss women.

35   The chymists are a strange class of mortals, impelled by an almost insane impulse to seek their pleasure among smoke and vapour, soot and flame, poisons and poverty; yet among all these evils I seem to live so sweetly that I may die if I would change places with the Persian King.
Physica Subterranea on title pages of *Brighter Biochemistry* in the 1930s, the journal of the Biochemical Laboratory in Cambridge.

36   A custome lothsome to the eye, hatefull to the Nose, harmfull to the braine, daungerous to the Lungs, and in the blacke stinking fume thereof, neerest resembling the horrible Stigian smoke of the pit that is bottomlesse.
A Counterblast to Tobacco by R B 1604 (London)

37   Dear Dr. Marx, If you are unable to produce the m/s in time we will be forced to have this book on Capital written by someone else.
[Attributed]

38   An education enables you to earn more than an educator.
In Hans Gaffron *Resistance to Knowledge* 1970 (San Diego, CA: Salk Institute)

39   *Fac ita.*
Do it so.
[Expression of early German and Italian algebraicists]

40   First baseball umpire: 'Balls and strikes, I call them as I sees them.' Second umpire: 'Balls and strikes, I call them as they are.' Third umpire: 'Balls and strikes, they ain't nothing until I call them.'
In A Rapoport *Strategy and Conscience* 1969 (New York: Schocken) p 194

41   *Gnothi seauton.*
Know thyself.
[From the Temple of Apollo at Delphi]
Pausanias 10.24.1; *Juvenal* 11.27

42   God is not dead: He is alive and well and working on a much less ambitious project.
Grafitto, London, 1975

43   Half of the secret of resistance to disease is cleanliness; the other half is dirtiness.

44   *Hic locus est ubi mors gaudet succurere vitae.*
This place is where death rejoices to come to the aid of life.
[In the anatomical dissection theatre of the University of Bologna]
In *Scientific American* March 1983 p 121

45   How many people work in your department ?—About a quarter.

46   [The human machine cannot even stand with its control system (brain) switched off, but] The centipede does not fall over even when it is dead.
Chinese proverb

47   I am accustomed to my deafness,
To my dentures I am resigned,
I can manage my bifocals,
But, oh, how I miss my mind!
In [Lord] Douglas Home *Asian Affairs* 1977 **8** pt II 247

48   If a research project is not worth doing at all, it is not worth doing properly.
J. Irreproducible Results 1961 **9** 43. Reprinted in A Kohn *Important Laws in Science*
(a review)

49   I hear, and I forget
I see, and I remember
I do, and I understand.

50   I, Lieutenant Orlov, Yuri Aleksandrovich, in the name of the Warsaw Pact
armies, order that work be stopped as of 1 pm August 22. All workers and
members are to leave the building until further notice.
Notice posted on the building of the Czechoslovak Academy of Sciences by the Soviet
Army, August 1968. In *Science* 1968 **161** 1326

51   In correspondence with the needs of society the state will support the
planned development of science and the training of scientific personnel and
will organise the introduction of the results of scientific research into the
national economy and other spheres of life.
Article 26 of the Soviet Constitution. Likhtenshtein, II, p 3

52   Is the present severe winter due to the impiety in the city?
[4th century BC lead tablet at Dodona, northern Greece. A question to the Oracle.]
In *Phil. Trans. R. Soc.* A 1990 **330** 650, from H W Parke *The Oracles of Zeus:*
Dodona, Olympia, Ammon 1967 (Oxford: Blackwell) pp 261–2

53   It is an established rule of the Royal Society ... never to give their opinion,
as a Body, upon any subject, either of Nature or Art, that comes before
them.
Advertisement in each issue of the *Philosophical Transactions of the Royal Society*
up to the 1950s in Nigel Calder *Technopolis* 1969 p 15

54   'It is better to be a crystal and be broken than to remain perfect as a tile
upon the housetop.'
In R L Stevenson *Familiar Stories of Men and Books. The Works of R L Steven-*
son vol III, Swanston Edition p 140 [quoted by Kusakabe to Yoshida on his way to
execution]

55   It works better if you switch it on.

56   Jesus saves—but Millikan gets the credit.
Graffito at California Institute of Technology, of which R A Millikan was then Director.
In *The Guardian* 28 September 1989

57   *Koń jaki jest, każdy widzi.*
What a horse is, is evident to everyone.
[Definition of a horse in the first Polish encyclopaedia.]
In Alfred Majewicz (Poznan) *Bul. Inst. for the Study of N. Eurasian Cultures* 1987
(Hokkaido: Hoppo Bunka Kenkyu) **18** 279

58   *Kuro Yagi-san kara o-tegami tsuita*
Shiro Yagi-san tara yomazu ni tabeta.
Shikata ga nai no de o-tegami kaita:
'Sakki no tegami no go-youji nani?'
Shiro Yagi-san kara o-tegami tsuita,
Kuro Yagi-san tara yomazu ni tabeta. ...
A letter arrived from Black Goat. White Goat ate the letter before reading
it. He could do nothing but write a letter: 'What were the contents of your
last letter?'

A letter arrived from White Goat. Black Goat ate the letter before reading it ...
Japanese children's song

59   Laws of Thermodynamics:
1. You cannot win.
2. You cannot break even.
3. You cannot get out of the game.

60   Love me, love my dogma.

61   *Magna opera domini exquisita in omnes voluntates eius.*
The works of the Lord are great; sought out of all those that have pleasure therein.
[Over the gateways of the Cavendish Laboratory, Cambridge]

62   Man occasionally stumbles on the truth, but then just picks himself up and hurries on regardless.

63   May all the gods curse him with a curse which cannot be relieved, terrible and merciless, as long as he lives, may they let his name, his seed, be carried off from the land, may they put his flesh into a dog's mouth.
To losers of tablets in the library of Assurbanipal, King of Assyria, 7th century BC

64   Me, I'm a morphologist—I sleep most of the time.
[morphe=shape, but Morpheus, the God of Sleep, is properly called 'the moulder', from the shapes he calls up before the sleeper.]

65   The most important item of research equipment is the Boeing 707.

66   The most powerful antigen in human biology is a new idea.

67   Multiplication is vexation; Division is as bad; The Rule of Three perplexes me, And fractions drive me mad.

68   My karma has run over my dogma.
Grafitto, Kentish Town, 1982

69   Nature requires five, Custom allows seven, Idleness takes nine, And wickedness eleven.
[hours in bed]
Mother Goose

70   A neurotic is a man who builds castles in the air; a psychotic is one who tries to live in them, and the psychiatrist is the man who collects the rent.

71   Never go to sea with two chronometers; take one or three.
In Frederick P Brooks *The Mythical Man-month* 1975 (San Frnacisco: Addison-Wesley) p 65

72   *Nihil est in intellectu quod non prius fuerit in sensu.*
There is nothing in the mind that has not been previously in the senses.

73   Not only are our highest Ministers of State ignorant of science, but the same defect runs through almost all the public departments of the Civil Service. It is nearly universal in the House of Commons, and is shared by the general public, including a large proportion of those engaged in industrial and commercial enterprise.
[Memorandum leading to the formation of the Neglect of Science Committee]
The Times 2 February 1916

74   The number you have dialed is imaginary. Please multiply by *i* and dial
      again.
      MIT telephone exchange (attributed)

75   Only we can prevent forests.
      [Slogan hung up in the headquarters of the US Air Force distributing defoliants in
      Vietnam]
      In R Clarke *We all fall down* 1968 (Harmondsworth: Pelican) p 117

76   *Pater semper incertus, mater certissima est.*
      The father is always in doubt, but the mother is most certain.

77   The Philosophy of Princes is to dive into the secrets of men, leaving the
      secrets of nature to those that have spare time.
      In George Herbert *George Herbert* 1659 p 373

78   *Pour moi, faire la conquête de la nature c'est la même chose que faire la
      conquête d'une femme.*
      [Contemporary scientist whose name Bronowski declines to quote] In J Bronowski
      Magic, Science and Civilisation 1978 (New York: Columbia University Press) p 40
      [Bronowski adds 'That's very much a 1500, white magic view; a little out of date, but
      that's how the French are.']

79   Rest in peace. The mistake shall not be repeated.
      [Cenotaph in Hiroshima]
      Scientific American June 1968 p 132

80   *Savoir*
      Savoir faire
      Faire
      Faire savoir.
      To know
      To know what to do
      To do it
      To make it known.
      Department of Physics, Trinity College, Dublin

81   *Shui sheng mei you gu jin.*
      The sound of water has neither past nor present.

82   '*Simplex sigillum veri*
      Cut causes, be merry
      Slash 'em and dock 'em'
      Said William of Ockham
      Wiping his razor
      On the sleeve of his blazer.
      In *The Times Literary Supplement* 18 June 1981, p 688

83   Sis, I have found out that there is no Santa Claus, and when I'm a little
      older, I'm going to look into this stork business, too.
      Perspectives in Biology and Medicine Autumn 1972 p 89

84   So, He was a great teacher. But what did he publish?
      [From an Israeli academic]

85   The strangest things happen in history. I once had an old coat, which I
      gave to my father; he used it for gardening and when it was worn out he

gave it to his father, who gave it to his father again, and so on until it was lost at the Battle of Stamford Bridge.
Old Danish anecdote in O H Iverson *Chalones* ed J C Houck, 1976 (Amsterdam: North-Holland) p 40

86   *Tam veneshlam shwach le El bore olam.*
Finished and completed, praise to the Lord, creator of the world.
[Traditional ending to a Hebrew book]

87   Thank God for what *Nature* [the periodical] allows—namely, the possibility of not revealing anything of one's self, if one so chooses.
Anna Brito in June Goodfield *An Imagined World* 1981 (Harmondsworth: Penguin) p 236

88   This stone commemorates the exploit of William Webb Ellis who, with a fine disregard for the rules of football as played in his time, first took the ball in his arms and ran with it, thus originating the distinctive feature of the Rugby game. AD 1823.
[Perhaps also characteristic of the English school of physics.]
(Rugby School, England)

89   To be a Greek was to seek to know, to know the primordial substance of matter; to know the meaning of number, to know the world as a rational whole.
In T L Heath *The Legacy of Greece* (Oxford: Oxford University Press) p 97

90   To be or not to be,
To be is to do
To do is to be
Doobie, Doobie, Doobie, Do.
The statistical physicist's ditty, with increasing entropy. *Nature* 1988 **336** 119

91   The trouble with facts is that there are so many of them.

92   Twinkle, twinkle little star,
I don't wonder what you are,
For by spectroscopic ken
I know that you are hydrogen.
In D Bush *Science and English Poetry* 1950 (Oxford: Oxford University Press) p 143

93   The Vice-Chancellor of the University of Cambridge presents his compliments to the Directors of the Eastern Counties Railway and begs to inform them that he has learnt with regret that it is the intention of the Directors of the Eastern Counties Railway to run excursion trains to Cambridge on the Lord's Day; with the object of attracting foreigners and undesirable characters to the University of Cambridge on that sacred day. The Vice-Chancellor of the University of Cambridge wishes to point out to the Directors of the Eastern Counties Railway that such a proceeding would be as displeasing to Almighty God as it is to the Vice-Chancellor of the University of Cambridge. (1844)
In Lawrence Wright *Clockwork and Man* 1968 (London: Elek) p 147

94   What is matter?—Never mind.
What is mind?—No matter.
Thomas Hewitt Key (1799–1875) in *Punch* 14 July 1855

95   When all else fails, read the instructions.

96   *Warum einfach, wenn es kompliziert geht?*
        Why bother to make it elegant if it already works?

97   We know that the magnet loves the lodestone, but we do not know whether
        the lodestone also loves the magnet or is attracted to it against its will.
        [Arab physicist of the 12th century]
        In D Gabor *Inventing the Future* 1964 (Harmondsworth: Penguin) p 187

98   The Zuñi nuclear family consists of a father, a mother, two children and
        an anthropologist.
        In G M Weinberg *An Introduction to General Systems Thinking* 1975 (New York:
        Wiley) p 119

[Saint] Anselm ca 1100

99   *Credo ut intelligam.*
        I believe so that I may understand.

Guillaume Apollinaire 1880–1918

100   [Of the Cubists] ... we who are constantly fighting along the frontiers of
        the infinite and of the future.
        Probably C Grey *Cubist Aesthetic Theories* (Baltimore, MD: Johns Hopkins Press)

[Saint] Thomas Aquinas ca 1225–1274

101   Practical sciences proceed by building up; theoretical sciences by resolving
        into components.
        Commentary on the Ethics I, 3

102   If, therefore, angels are not composed of matter and form, as was said
        above, it follows that it would be impossible to have two angels of the same
        species .... The motion of an angel can be continuous or discontinuous as
        it wishes ... And thus an angel can be at one instant in one place, and at
        another instant in another place, not existing at any intermediate time.
        [The Pauli exclusion principle?]
        Summa Theologica I, 50, 4

François Arago 1786–1853

103   *Connaître, découvrir, communiquer—telle est la destinée d'un savant.*
        To get to know, to discover, to publish—this is the destiny of a scientist.

Michael A Arbib 1940–

104   In the beginning was the word
        and by the mutations came the gene.
        [WORD–WORE–GORE–GONE–GENE]
        Towards a Theoretical Biology vol 2, ed C H Waddington,
        1969 (Edinburgh: Edinburgh University Press) p 323

Archilochos of Paros ca 714–676 BC

105   The fox knows many things, but the hedgehog one big thing.
In E Diehl *Anthologia lyrica Graeca* 1949–52 no 103

Archimedes ca 287–212 BC

106   Any solid lighter than a fluid will, if placed in the fluid, be so far immersed
that the weight of the solid will be equal to the weight of the fluid displaced.
On Floating Bodies I, prop. 5

107   Archimedes to Eratosthenes greeting. ... certain things first became clear
to me by a mechanical method, although they had to be demonstrated by
geometry afterwards because their investigation by the said method did not
furnish an actual demonstration. But it is of course easier, when we have
previously acquired, by the method, some knowledge of the questions, to
supply the proof than it is to find it without any previous knowledge.
The Method in *The Works of Archimedes* transl T L Heath 1912 (Cambridge: Cambridge University Press)

108   *Eureka, Eureka.*
I have found [it]!
Vitruvius Pollio *De Architectura* ix, 215

109   Give me a place to stand, and I will move the earth ....
On the lever in Pappus *Synagoge* ed F Hultsch, 1868 Berlin **VIII** 10, xi

Robert Ardrey

110   Man, the bad weather animal.
African Genesis 1961 Fontana-Collins, ch 9

Aristophanes ca 444–ca 380 BC

111   First listen, my friend, and then you may shriek and bluster.
Ecclesiazousae 588f

112   *Socrates:* Suppose you are arrested for a debt,
We'll say five talents, how will you contrive
to cancel at a stroke both debt and writ?
... *Strepsiades:* I've hit the nail
That does the deed, and so you will confess.
Socrates: Out with it.
Strepsiades: Good chance but you have noted
A pretty toy, a trinket in the shops,
Which being rightly held produceth fire
From things combustible—
Socrates: A burning glass, Vulgarly called—
Strepsiades: You are right; 'tis so.
Socrates: Proceed.
Strepsiades: Put the case now your whoreson baliff comes,
Shows me his writ
—I, standing thus, d'ye mark me,
In the sun's stream, measuring my distance, guide
My focus to a point upon his writ,
And off it goes *in fumo.*
The Clouds transl T Mitchell, 1911 (London: Dent)

113   *Suidas:* Antiphon, an Athenian, an interpreter of signs, an epic poet and a sophist. He was called a word-cook.

114   *Meton:* With the straight ruler I set to work
To make the circle four-cornered.
[Allusion to the problem of the quadrature of the circle—of constructing, with only ruler and compasses, a square with the same area as a given circle.]

Aristotle 384–322 BC

115   If every tool, when ordered, or even of its own accord, could do the work that befits it, just as the creations of Daedalus moved of themselves .... If the weavers' shuttles were to weave of themselves, then there would be no need either of apprentices for the master workers or of slaves for the lords.
Atheniensium respublica transl F G Kenyon, 1920

116   If one way be better than another, that you may be sure is Nature's way.
Nicomanachean Ethics 1099B, 23

117   If this is a straight line [showing his audience a straight line drawn by a ruler], then it necessarily ensues that the sum of the angles of the triangle is equal to two right angles; and conversely, if the sum is not equal to two right angles, then neither is the triangle rectilinear.
Physica

118   It is not once nor twice but times without number that the same ideas make their appearance in the world.
On the Heavens in T L Heath *Manual of Greek Mathematics* 1931 (Oxford: Oxford University Press) p 205

119   Now that practical skills have developed enough to provide adequately for material needs, one of those sciences which are not devoted to utilitarian ends [mathematics] has been able to arise in Egypt, the priestly caste there having the leisure necessary for disinterested research.
Metaphysics I-981b

120   Speech is the representation of the mind, and writing is the representation of speech.
De interpretatione 1

121   The whole is more than the sum of the parts.
Metaphysica 1045a, 10f

122   [Quoting Agathon] Chance is beloved of Art, and Art of Chance ....
Scientific American August 1969, p 132

123   What we have to learn to do, we learn by doing.
Ethica Nichomachea II

124   The body is most fully developed [at] from thirty to thirty-five years of age, the mind at about forty-nine.
Rhetoric II 14

Louis Armand 1905–

125   *Quand le standard de vie s'élève, le rapport*
(condition de l'homme)/(condition de la femme) tend vers 1.
When the standard of living is improved, the ratio
(condition of man)/(condition of woman) tends to one.
Plaidoyer pour l'Avenir La Technique, Moyen de Civilisation, Calmann-Levy

Neil A Armstrong 1930–

126   One small step for man, one big step for mankind.
[21 July 1969. First words on stepping on to the Moon] *Nature* 1974 **250** p 451

127   That's one small step for a man, one giant leap for mankind.
[Official version]

Thomas Arnold 1795–1842

128   Rather than have Physical Science the principal thing in my son's mind, I would rather have him think that the Sun went round the Earth, and that the Stars were merely spangles set in a bright blue firmament.
Penguin Dictionary of Quotations

Lev Andreevich Artsimovich 1909–1973

129   The most important thing for a genius is to be born at the right time.

Roger Ascham 1515–1568

130   Mark all Mathematical heads which be wholly and only bent on these sciences, how solitary they be themselves, how unfit to live with others, how unapt to serve the world.
In E G R Taylor *Mathematical Practitioners of Tudor and Stuart England* 1954 (Cambridge: Cambridge University Press)

[Lord] Eric Ashby 1904–

131   The habit of apprehending a technology in its completeness. this is the essence of technological humanism, and this is what we should expect education in higher technology to achieve. I believe it could be achieved by making specialist studies the core around which are grouped liberal studies which are relevant to these specialist studies. But they must be relevant; the path to culture should be through a man's specialism, not by-passing it . . . . A student who can weave his technology into the fabric of society can claim to have a liberal education; a student who cannot weave his technology into the fabric of society cannot claim even to be a good technologist.
Technology and the Academics 1958 (London: Macmillan)

132   The notion that the 'balance of nature' is delicately poised and easily upset is nonsense. Nature is extraordinarily tough and resilient, interlaced with checks and balances, with an astonishing capacity for recovering from disturbances in equilibrium. The formula for survival is not power; it is symbiosis.
Encounter March 1976, p 16

133   In the university, the specialist and analyst is king. But resolution of problems in society generally is not to be found in a single discipline . . . . In society the non-specialist and synthesiser is king.
The Sociology of Science ed P Halmers, 1972, Sociology Review Monographs no 18

William Thomas Astbury 1898–1961

134   [Molecular biology (note)] is concerned particularly with the forms of biological molecules and with the evolution, exploitation and ramification of these forms in the ascent to higher and higher levels of organisation. Molecular biology is predominantly three-dimensional and structural—which does

not mean, however, that it is merely a refinement of morphology. It must
at the same time enquire into genesis and function.
In *Science* 1968 **160** 390. From *Harvey Lectures, 1950–51* 1952 (Springfield, IL:
Thomas) p 3.
[Note: 'molecular biologist' *Ann. Rev. Biochem.* 1939 **8** 113, 125; also 28 September
1951, *Harvey Lecture Series, 1951* **46** 3; *Nature* 1961 **190** 1124

Peter William Atkins 1940–

135   Freewill is merely the ability to decide, and the ability to decide is nothing
other than the organised interplay of shifts of atoms responding to freedom
as chance first endows them with energy to explore, and then traps them
in new arrangements as their energy leaps naturally and randomly away.
Even freewill is ultimately corruption.
The Creation 1981

John Aubrey 1626–1697

136   ... 'twas held a strange presumption for a man to attempt improvement
of any knowledge whatsoever; they thought it not fit to be wiser than their
fathers and not good manners to be wiser than their neighbours; and a sin
to search into the ways of nature.
In Michael Hunter *John Aubrey and the Realm of Learning* 1975 (London: Duckworth)

137   [Of Thomas Hobbes, in 1629] He was 40 years old before he looked on geometry; which happened accidentally. Being in a gentleman's library, Euclid's
Elements lay open, and 'twas the 47 *El. libri* I [Pythagoras' Theorem]. He
read the propostition. 'By God', sayd he. (He would now and then swear,
by way of emphasis.) 'By God', sayd he, 'this is impossible:' So he reads
the demonstration of it, which referred him back to such a proposition;
which proposition he read. That referred him back to another, which he
also read. *Et sic deinceps*, that at last he was demonstratively convinced of
that trueth. This made him in love with geometry.
Brief lives ed O L Dick, 1960 (Oxford: Oxford University Press) p 604

138   Old Customs and old wives fables are grosse things, but yet ought not to be
buried in Oblivion; ther may be some trueth and usefulnesse to be picked
out of them, besides, 'tis a pleasure to consider the errors that enveloped
former ages: as also the present.
Aubrey's Brief Lives ed O L Dick, 1949 (London: Secker and Warburg) p lxv

139   [Of Robert Hooke (1635–1703)] Mr Hooke sent, in his next letter [to Sir
Isaac Newton] the whole of his Hypothesis, scil. that the gravitation was
reciprocal to the square of the distance: .... This is the greatest Discovery
in Nature that ever was since the World's creation. It was never so much
as hinted by any man before. I wish he had writt plainer, and afforded a
little more paper.
Aubrey's Brief Lives ed O L Dick, 1949 (London: Secker and Warburg) p 166–7

140   As a supplement to Sir William Petty's excellent 'Advice of Mechanics' I
would have boys that are mechanically given, to learn the tin-man's trade
which will be greatly useful for making of models; 'tis easily learned and
two or three tools will serve their turn.
Aubrey on Education ed J E Stephens, 1971 (London: Routledge, Kegan Paul) p 111

141   [Of Sir Charles Cavendish, mathematician (1591–1654)] He had collected in
Italie, France, etc., with no small charge, as many Manuscript Mathematical

bookes as filled a Hoggeshead, which he intended to have printed; which, if
he had lived to have done, the growth of Mathematical learning had been
30 years or more forwarder than 'tis. But he died of the Scurvey, contracted
by hard study ... and left his wife Executrix, who sold this incomparable
Collection aforesaid, by weight to the past-board makers for Wast-paper.
A good Caution to those that have good MSS. to take care to see them
printed in their life-times.
Aubrey's Brief Lives ed O L Dick, 1949 (London: Secker and Warburg) p 58

Wystan Hugh Auden 1907–1973

142   But he would have us remember most of all
to be enthusiastic over the night
Not only for the sense of wonder
It alone has to offer, but also
Because it needs our love: for with sad eyes
Its delectable creatures look up and beg
Us dumbly to ask them to follow;
They are exiles who long for the future.
In Memory of Sigmund Freud 1951 (New Haven, CT: Yale University Press)

143   He was found by the Bureau of Statistics to be
One against whom there was no official complaint, ...
Another Time 1940 in *The Unknown Citizen*

144   How happy the lot of the mathematician. He is judged solely by his peers,
and the standard is so high that no colleague or rival can ever win a reputation he does not deserve.
The Dyer's Hand 1948 (London: Faber & Faber) p 15

145   Learning screams in the narrow gate where
Events are traded with time but cannot
Tell what logic must and must not leave to fate
Or what laws we are permitted to obey;
There are no birds, the predatory
Glaciers glitter in the chilly evening.

146   Love requires an Object
But this varies so much
Almost, I imagine,
Anything will do:
When I was a child, I
Loved a pumping-engine,
Thought it every bit as
Beautiful as you.
Heavy Date in *Collected Shorter Poems, 1927–1957* 1966 (London: Faber & Faber)
p 153

147   Of course, Behaviourism 'works'. So does torture. Give me a no-nonsense,
down-to-earth behaviourist, a few drugs, and simple electrical appliances,
and in six months I will have him reciting the Athanasian Creed in public.
A Certain World 1970 (London: Faber & Faber) p 33

148   Those who will not reason
Perish in the act:

Those who will not act
Perish for that reason.
In *Collected Shorter Poems, 1927–1957* 1966 (London: Faber & Faber) p 42

149   Thou shalt not answer questionnaires
Or quizzes upon world affairs,
Nor with compliance Take any test.
Thou shalt not sit with statisticians nor commit
A social science.
Under which lyre in *Collected Poems of W H Auden* (London: Faber & Faber)

150   To the man-in-the-street, who, I'm sorry to say
Is a keen observer of life,
The word intellectual suggests straight away
A man who's untrue to his wife.
Note on Intellectuals in *Collected Shorter Poems, 1927–1957* 1966 (London: Faber & Faber)

151   Tomorrow, perhaps, the future: the research on fatigue
And the movements of packers; the gradual exploring of all the
Octaves of radiation;
Spain in *Another Time* 1940 (London: Faber & Faber)

152   When I find myself in the company of scientists, I feel like a shabby curate
who has strayed by mistake into a drawing room full of dukes.
The Dyer's Hand and other Essays 1962 (New York: Random House) p 81

153   No tyrant ever fears
His geologists or his engineers.
City Without Walls 1969 (London: Faber & Faber) p 58

154   ... the One for Whom all
enantiomorphs
are superposable, yet
Who numbers each particle
by its Proper Name.
Epithalmium (1965) in *City Without Walls* 1969 (London: Faber & Faber) p 22

155   Nature rewards perilous leaps. The prudent atom insists upon its safety
now.
The Age of Anxiety in University of London Bulletin, no 37, December 1976, p 2

156   Science, like Art, is fun, a playing with truths, and no game
should ever pretend to slay the heavy-lidded riddle,
What is the Good Life?
Unpredictable but Providential for Loren Eisley

Stan Augarten

157   Computers are composed of nothing more than logic gates stretched out to
the horizon in a vast numerical irrigation system.
State of the Art: A Photographic History of the Integrated Circuit (New York: Ticknor
and Fields) in *Scientific American* January 1984, p 19

Pierre Auger 1899–

158   Holophrase—Multisyllabic words designating precisely a very complex situation but entirely *sui generis* and not resolvable; that is, containing neither

roots not affixes used elsewhere. The best known example is *mamihlapinata-pai* which in Tierra des Fuego designates the situation in which two persons look at each other, each hoping to see the other undertake an action which both wish for without being disposed to take the initiative.

[The main use of this word is in describing the US/USSR talks on the limitation of strategic arms]

The Regime of Castes in the Populations of Ideas in *Diogenes* 1958 **22** 42

[Saint] Augustine [of Hippo] 354–430

159 *Angelus non potest peccare; homo potest non peccare.*
An angel cannot sin; a man can choose not to sin.

160 The good Christian should beware of mathematicians, [astrologers] and all those who make empty prophecies. The danger already exists that the mathematicians have made a covenant with the devil to darken the spirit and to confine man in the bonds of Hell.
In M Kline *Mathematics in Western Culture* 1953, p 3 (*De Genesi ad Litteram*, book II, xvii, 37)

161 If I am given a sign [formula], and I am ignorant of its meaning, it cannot teach me anything, but if I already know it what does the sign teach me?
De Magistro ch X, 23

162 *Nisi credideritis, non intelligitis.*
If you don't believe it, you won't understand it.
De Libero Arbitrio

163 There is another form of temptation even more fraught with danger. This is the disease of curiosity .... It is this which drives us on to try to discover the secrets of nature, those secrets which are beyond our understanding, which can avail us nothing and which men should not wish to learn.
In C Sagan *Dragons of Eden* p 236

164 The world was made, not in time, but simultaneously with time. There was no time before the world.
Confessions

Antoninus Marcus Aurelius 121–180

165 Everything that happens, happens as it should, and if you observe carefully, you will find this to be so.
Meditations IV, 10

166 That which is not good for the bee-hive, cannot be good for the bees.
[Like General Motors and the USA]
Meditations VI, 49

167 The world is either the effect of cause or of chance. If the latter, it is a world for all that, that is to say, it is a regular and beautiful structure.
Meditations IV, 22

Avicenna [Ibn Sina] 980–1037

168 *non turpe est medico, cum de rebus veneris loquitur, de delectatione mulieris coeuntis: quoniam sunt ex causis, quibus pervenitur ad generationem.*
Writing about erotics is a perfectly respectable function of medicine, and about the way to make the woman enjoy sex; these are an important part of reproductive physiology.
In Alex Comfort *Sex in Society*. From the *Qanun* [Canon of Medicine]

Ruby Mildred Ayres 1883–1953

169   First I fix the price. Then I fix the title. Then I write the book.
      In her obituary in *The Times*. Also in W Gratzer *Nature* 1983 **302**, 165

Michael Ayrton 1921–1975

170   *Daedalus:* Each man's life is a labyrinth at the centre of which lies his death,
      and even after death it may be that he passes through a final maze before it
      is all ended for him. Within the great maze of a man's life are many smaller
      ones, each seemingly complete in itself, and in passing through each one he
      dies in part, for in each he leaves behind him a part of his life and it lies
      dead behind him. It is a paradox of the labyrinth that its centre appears
      to be the way to freedom.
      The Maze Maker 1967

171   I do not believe that it is possible to create living art out of anything but
      the direct visual experience of nature, combined with the heritage of a
      tradition, unless it be by the practice of magic ritual.
      Golden Sections 1957 (London: Methuen) p 80

172   *Daedalus:* I am not impious nor insensitive to the visions of poets and
      other sacred persons, but in general they are less observant than they think,
      suffering as they do from revelation, which blinds them. I accept that poets
      celebrate important things such as honour and beauty and birth and valour
      and man's relations with the gods but, when you come down to it, what
      they most celebrate are heroes, which is not surprising. Poets have much
      in common with heroes. They are neither of them aware of the world, of its
      true appearance nor its real consequence, its structure nor its marvellous
      imperfection. They are blind to that and because my methods of gaining
      experience have been observation, deduction and experiment, I have been
      no worse off and much better instructed than any poets or heroes known
      to me. In fact since I am not beset by my own personality I am better off.
      I prefer cognition to revelation and in my view the valiant act is to live
      as long and as fully as possible, but then I make things which take time.
      Honour lasts longer if it is gained by patience rather than by some noble
      gesture rapidly made.
      The Maze Maker 1967 (London: Longmans) p 117

Azerbaijani wall poster

173   Without sanitary culture, there would be no culture at all.
      [Seen on a banner outside the Academy of Sciences in Baku, 1962]

Charles Babbage 1792–1871

1   Errors using inadequate data are much less than those using no data at all.

2   Remember that accumulated knowledge, like accumulated capital, increases
at compound interest: but it differs from the accumulation of capital in
this; that the increase of knowledge produces a more rapid rate of progress,
whilst the accumulation of capital leads to a lower rate of interest. Capital thus checks its own accumulation: knowledge thus accelerates its own
advance. Each generation, therefore, to deserve comparison with its prede-
cessor, is bound to add much more largely to the common stock than that
which it immediately succeeds.
The Exposition of 1851 1851 (London: Murray) p 25

3   That the master manufacturer, by dividing the work to be executed into
different processes, each requiring different degrees of skill and strength, can
purchase exactly that precise quantity of both which is necessary for each
process; whereas, if the whole work were executed by one workman, that
person must possess sufficient skill to perform the most difficult, and sufficient strength to execute the most laborious, of the various operations...
. When (from the peculiar nature of the produce of each manufactory) the
number of processes into which it is most advantageous to divide it is ascertained, as well as the number of individuals to be employed, then all
other manufactories which do not employ a direct multiple of this number, will produce the article at greater cost. This principle ought always
to be kept in view in great establishments, although it is quite impossible,
even with the best system of the division of labour, to carry it rigidly into
execution.... .
On the Economy of Machinery and Manufacturers London, 1832

4   The whole of the developments and operations of analysis are now capable
of being executed by machinery.... As soon as an Analytical Engine exists,
it will necessarily guide the future course of science.
Passages from the Life of a Philosopher 1864 (London: Longman)

Gaston Bachélard 1884–1962

5   *L'observation scientifique est toujours une observation polémique; elle
confirme ou infirme une thèse antérieure; un schéma prealable, un plan
d'observation; elle montre en démontrant; elle hiérarchise les apparences;
elle trancende l'immédiat; elle reconstruit le réel après avoir reconstruit ses
schémas.*
A scientific observation is always a committed observation. It confirms or
denies one's preconceptions, one's first ideas, one's plan of observation. It
shows by demonstration. It structures the phenomena. It transcends what
is close at hand. It reconstructs the real after having reconstructed its
representation.
La nouvel esprit scientifique 1934, vol I (Presses Univ. de France)

6   *Balzac disait que les célébataires remplacent les sentiments par des habi-
tudes. De même, les professeurs remplacent les découvertes par des leçons.*
Balzac said that bachelors replace feelings by habits. In the same way,
academics replace research by teaching.
La formation de l'ésprit scientifique 1938 (Paris: Vrin) p 247

[Sir] Francis Bacon [Lord Verulam] 1561-1626

7   And as for Mixed Mathematics, I may only make this prediction, that there cannot fail to be more kinds of them, as nature grows further disclosed.
Advancement of Learning book 2; *De Augmentis* book 3

8   Books must follow sciences, and not sciences books.
Verulamiana London, 1803 p 15

9   The End of our Foundation is the knowledge of Causes, and the secret motions of things; and the enlarging of the bounds of Human Empire, to the effecting of all things possible.
New Atlantis 1626

10   For man being the minister and interpreter of nature, acts and understands so far as he has observed of the order, the works and mind of nature, and can proceed no further; for no power is able to loose or break the chain of causes, nor is nature to be conquered but by submission: whence those twin intentions, human knowledge and human power, are really coincident; and the greatest hindrance to works is the ignorance of causes.
Towards the end of the Preface to *The Great Instauration*

11   He that will not apply new remedies must expect new evils; for time is the greatest innovator.
On innovations Essays

12   The human Intellect, in those things which have once pleased it (either because they are generally received and believed, or because they suit the taste), brings everything else to support and agree with them; and though the weight and number of contradictory instances be superior, still it either overlooks or despises them, or gets rid of them by creating distinctions, not without great and injurious prejudice, that the authority of these previous conclusions may be maintained inviolate. And so he made a good answer, who, when he was shown, hung up in a temple, the votive tablets of those who had fulfilled their vows after escaping from shipwreck, and was pressed with the question, 'Did he not then recognize the will of the gods' asked, in his turn, 'But where are the pictures of those who have perished, notwithstanding their vows?' The same holds true of almost every superstition—as astrology, dreams, omens, judgments, and the like—wherein men, pleased with such vanities, attend to those events which are fulfilments; but neglect and pass over the instances where they fail (though this is much more frequently the case).
Novum Organum 1620, book 1, aphorism XLVI

13   It is well to observe the force and virtue and consequence of discoveries, and these are to be seen nowhere more conspicuously than in printing, gunpowder and the magnet. For these three have changed the whole face and state of things throughout the world, ... in so much that no empire, no sect, no star seems to have exerted greater power and influence in human affairs than these mechanical discoveries.
Novum Organum 1620

14   It would be the height of folly—and self-defeating—to think that things never heretofore done can be accomplished without means never heretofore tried.
Novum Organum 1620, book I, sec. vi

15   *Nam et ipsa scientia potestas est.*
For knowledge itself is power.
Religious Meditations, Of Heresies

16   *Natura non nisi parendo vincitur.*
Nature, to be commanded, must be obeyed.
Novum Organum 1620

17   Never any knowledge was delivered in the same order it was invented.

18   Reading maketh a full man; conference a ready man and writing an exact man.
[Bacon maketh a fat man—graffito]
Of Studies Essay 50

19   Truth comes out of error more readily than out of confusion.
Novum Organum 1620

20   Histories make men wise; poets, witty; the mathematics subtile; natural philosophy, deep; moral, grave; logic and rhetoric, able to contend.
Of Studies Essay 50

21   The ill and unfit choice of words wonderfully obstructs the understanding.
Novum Organum 1620, Aphorism 43

Roger Bacon ca 1214–1294

22   *Et harum scientarum porta et clavis est Mathematica.*
Mathematics is the door and the key to the sciences.
Opus Majus part 4 *Distinctia Prima* cap 1, 1267 transl Robert Belle Burke, 1928 (Philadelphia, PA: University of Pennsylvania Press)

23   For the things of this world cannot be made known without a knowledge of mathematics. For this is an assured fact in regard to celestial things, since two important sciences of mathematics treat of them, namely theoretical astrology and practical astrology. The first ... gives us definite information as to the number of the heavens and of the stars, whose size can be comprehended by means of instruments, and the shapes of all and their magnitudes and distances from the earth, and thicknesses and number, and greatness and smallness, ... It likewise treats of the size and shape of the habitable earth .... All this information is secured by means of instruments suitable for these purposes, and by tables and by canons .... For everything works through innate forces shown by lines, angles and figures.
Opus Majus transl Robert Belle Burke, 1928 (Philadelphia, PA: University of Pennsylvania Press)

24   *sed tamen salis petrae LURE MOPE CAN UBRE†, et sulphuris; et sic facies tonitrum, et corruseationem.*
In Erasmus Darwin *The Botanic Garden* London, 1791 p 25

---

†   Anagram of *carbonum pulvere*

Walter Bagehot 1826–1877

25   You may talk of the tyranny of Nero and Tiberias, but the real tyranny is the tyranny of your next-door neighbour. What espionnage of despotism comes to your door so effectively as the eye of the man who lives at your door? Public opinion is a permeating influence. It requires us to think other men's thoughts, to speak other men's words, to follow other men's habits.
The Works of Walter Bagehot 1889 (Hartford, CT: Traveller's Insurance Co)

Philip Noel Baker 1889–1982

26   The scientists speak with an authority which the ordinary citizen, the non-scientist, cannot challenge, and to which he is compelled to listen. Since they cannot hope for much help from the generals or the ministers, they must act for themselves, in a supreme endeavour to avert the mortal dangers which confront mankind.
Science and Disarmament, Impact 1965 **XV** no 4

T Baker

27   Rennie's conoidal triple-bladed screw.
Line of verse in *The Steam Engine* in *The Stuffed Owl*

Arthur [Earl of] Balfour 1848–1930

28   [Brother-in-law of Lord Rayleigh] ... science was the great instrument of social change, all the greater because its object is not change but knowledge and its silent appropriation of this dominant function, amid the din of political and religious strife, is the most vital of all the revolutions which have marked the development of modern civilisation.
Decadence 1908 (Cambridge: Cambridge University Press)

Walter William Rouse Ball 1850–1925

29   The manner of Demoivre's death has a certain interest for psychologists. Shortly before it, he declared that it was necessary for him to sleep some ten minutes or a quarter of an hour longer each day than the preceeding one: the day after he had thus reached a total of something over twenty-three hours he slept up to the limit of twenty-four hours, and then died in his sleep.
[Abraham de Moivre, 1667–1754]
History of Mathematics 1911 (London: Macmillan) p 394

Honoré de Balzac 1799–1850

30   No man should marry until he has studied anatomy and dissected at least one woman.
The Physiology of Marriage meditation v, aphorism 28

Nancy Banks-Smith

31   God invented sex so that we wouldn't take ourselves too seriously. Harumph he said, Immortality? I'll give you immortality! Now listen very carefully. This is what you do.
The Guardian 21 February 1990, p 38

[Sir] James Matthew Barrie 1860–1937

32   Those hateful persons called Original Researchers.
My Lady Nicotine Chapter 13

Gregory Bateson 1904–1980

33   Information is any difference that makes a difference.
In *Scientific American* September 1984, p 41

William Bateson 1861–1926

34   I would trust Shakespeare, but I would not trust a committee of Shake-
speares.
[The geneticist]
In J K Brierley *Biology and the Social Crisis* 1967 (London: Heinemann) p 75

Charles Pierre Baudelaire 1821–1867

35   *Un système est une espèce de damnation*
qui nous pousse à une abjuration perpetuelle;
il en faut toujours
un autre, et cette fatigue est un cruel
châtiment.
A system is a kind of damnation which drives us to perpetual abjuration.
We are forced to invent another, and this strain is a cruel punishment.
In J Culler *Structuralist Poetics* 1975 (London: Routledge & Kegan Paul) p 241

John Oliver Bayley 1925–

36   Being Polish is of course no joke, but Polish intellectuals tend to possess a
       detachment and a kind of sardonic equilibrium unpretended to by, say, their
       Irish equivalents. There is nothing peripheral about Poland; indeed there
       is something almost painfully central in her position as mediator between
       two cultures and powers of which she is the historical victim. To put it
       schematically, a French intellectual dances with an idea in a graceful minuet
       whose steps are self-determined by his language; a Russian is crushed by
       an idea and lies under it passively, like a man under a stone. Reaching
       out to France, Polish intelligence none the less remains clear-eyed about *la
       grande nation*, and can recognise the power of the Russian genius without
       succumbing to its *idées fixes*.
       Times Literary Supplement 2 December 1977, p 1419

Edmonde-Pierre Chanvot de Beauchêne 1748–1824

37   Science seldom renders men amiable; women, never.
       Maximes, réflexions et pensées diverses.

William S Beck 1923–

38   To understand culture one must have some understanding of biological
       thought.
       Modern Science and the Nature of Life 1957 (New York: Harcourt Brace)

39   Fields of learning are surrounded ultimately only by illusory boundaries—
       like the 'rooms' in a hall of mirrors. It is when the illusion is penetrated
       that progress takes place. ... Likewise science cannot be regarded as a
       thing apart, to be studied, admired or ignored. It is a vital part of our
       culture, our culture is part of it, it permeates our thinking, and its continued
       separateness from what is fondly called 'the humanities' is a preposterous
       practical joke on all thinking men.
       Modern Science and the Nature of Life 1957 (New York: Harcourt Brace)

40   The human problem, thus, would seem to be the art of avoiding the in-
       evitable consequences of a paternal creed: the surrender of intelligence in
       the purchase of emotional security—or, in a word, infantilism. For no hu-
       man institution, not culture, science, history, or the holy hegemony, is going
       to take care of man. Man, I'm afraid, is going to have to take care of himself.
       Modern Science and the Nature of Life 1957 (New York: Harcourt Brace)

Johann Beckmann 1739–1811

41   'Technology.'
       [Coinage of the word in his textbook of 1777]

Stafford Beer 1926–

42   *Absolutum obsoletum*—if it works it's out of date.
       Brain of the Firm 1972 (Harmondsworth: Penguin)

43   Cybernetics is mappable on to Marxist theory. Liberty must be a com-
       putable function.
       New Scientist 15 February 1973 p 347

44   *Man*: Hello, my boy. And what is your dog's name?
      Boy: I don't know. We call him Rover.
      New Scientist 3 October 1974 p.64

45   Our institutions are failing because they are disobeying laws of effective
      organisation which their administrators do not know about, to which indeed
      their cultural mind is closed, because they contend that there exists and
      can exist, no science competent to discover those laws.
      Designing Freedom 1974 (New York: Wiley) p 19

Belgian Notice

46   *Ne parler pas au wattman.*
      Do not talk to the tram-driver.
      [Thus immortalising James Watt]

Vissarion Grigorievich Belinskii 1811–1848

47   In science one must search for ideas. If there are no ideas, there is no science.
      A knowledge of facts is only valuable in so far as facts conceal ideas: facts
      without ideas are just the sweepings of the brain and the memory.
      Collected Works 1948 vol 2 (Moscow: OGIZ) p 348

Alexander Graham Bell 1847–1922

48   Mr Watson, come here, I want you!
      [The first telephone message]

Eric Temple Bell 1883–1960

49   The cowboys have a way of trussing up a steer or a pugnacious bronco
      which fixes the brute so that it can neither move nor think. This is the
      hog-tie, and it is what Euclid did to geometry.
      In R Crayshaw-Williams The Search for Truth p 191

Hillaire Belloc 1870–1953

50   Anyone of common mental and physical health can practise scientific re-
      search .... Anyone can try by patient experiment what happens if this or
      that substance be mixed in this or that proportion with some other under
      this or that condition. Anyone can vary the experiment in any number of
      ways. He that hits in this fashion on something novel and of use will have
      fame .... The fame will be the product of luck and industry. It will not be
      the product of special talent.
      Essays of a Catholic Layman in England 1931 (London: Sheed and Ward) p 226

51   ... but Scientists who ought to know
      Assure us [that] it must be so.
      Oh, let us never, never doubt
      What nobody is sure about.
      The Microbe in TIBS July 1978, p 164

52   Statistics are the triumph of the quantitative method, and the quantitative
      method is the victory of sterility and death.
      The Silence of the Sea

Nikolai Vassilevich Belov 1891–1982

53   [Obituary of J D Bernal] ... like a true Irishman, his last enthusiasm was for the laws of lawlessness.
Sov. J. Cryst. 1972 **17** 208–9

54   The exclusion of a five-fold axis for crystals, as well recognised, results from the impossibility of reconciling it (and axes of order greater than 6) with the 'lattice state' of crystalline matter. It would appear, then, that for small organisms the fivefold axis represents a distinctive instrument in their struggle for existence, acting as insurance against petrifaction, against crystallisation, in which the first step would be their 'capture' by a lattice.
Essays on structural crystallography XIII Mineral. Sbornik L'vov. Geol. Obschch. 1962 no 16 41

[Enoch] Arnold Bennett 1867–1931

55   [Of *Nature*] The writing of it is considerably inferior to the matter of it. ... I regard *Nature* as perhaps the most important weekly printed in English, far more important than any political weekly.
Evening Standard 20 November 1930 [*Nature* 1929 **126** 854; 1969 **224** 462

Henry Albert Bent 1926–

56   ... hell must be isothermal, for otherwise the resident engineers and physical chemists (of which there must be some) could set up a heat engine to run a refrigerator to cool off a portion of their surroundings to any desired temperature.
The Second Law 1965 (Oxford: Oxford University Press) p 313

57   The important point is not the bigness of Avogadro's number [6×10^{23} atoms/g atom] but the bigness of Avogadro.
[Avogadro consisted of some 10^{27} atoms]
The Second Law 1965 (Oxford: Oxford University Press)

Jeremy Bentham 1748–1832

58   O Logic: Born gatekeeper to the Temple of Science, victim of capricious destiny: doomed hitherto to be the drudge of pedants: come to the aid of thy master, Legislation.
Works ed J Browning, 1838–1843

Edmund Clerihew Bentley 1875–1956

59   Sir Humphrey Davy
Detested gravy.
He lived in the odium
Of having discovered Sodium.
Biography for Beginners 1925 (London: Werner Laurie)

Nicolas Berdiaeff 1874–1948

60   *Les utopies apparaissent comme bien plus réalisables qu'on ne le croyait autrefois. Et nous nous trouvons actuellement devant une question bien autrement angoissante: Comment éviter leur réalisation définitive ?*
Utopias now appear much more realizable than one used to think. We are now faced with a very different new worry: How to prevent their realization.
In Aldous Huxley *Brave New World* 1932 (London: Chatto & Windus)

Bernard Berenson 1865–1959

61   Our art has a fatal tendency to become science and we hardly possess a masterpiece which does not bear the marks of having been a battlefield for divided interests.
In Herbert Read *Icon and Idea* 1955 (London: Faber & Faber) p 101

Henri Bergson 1859–1941

62   *L'univers ... est une machine à faire des dieux.*
The universe is a device for making deities.
Les deux sources de la morale et de la religion 1932 (Paris: Presses Universitaires de France)

63   Calculation touches, at most, certain phenomena of organic destruction. Organic creation, on the contrary, the evolutionary phenomena which properly constitute life, we cannot in any way subject to a mathematical treatment.
Creative Evolution p 21

John Desmond Bernal 1901–1971

64   All that glisters may not be gold, but at least it contains free electrons.
[But consider the Golden Scarab Beetle which has a metallic lustre without metal]
Lecture at Birkbeck College, University of London, 1960

65   The beauty of life is, therefore, geometrical beauty of a type that Plato would have much appreciated.
The Origin of Life (London: Weidenfeld & Nicholson) 1967

66   But if capitalism had built up science as a productive force, the very character of the new mode of production was serving to make capitalism itself unnecessary.
Marx and Science 1952 (London: Lawrence and Wishart) p 39

67   The greater the man the more he is soaked in the atmosphere of his time; only thus can he get a wide enough grasp of it to be able to change substantially the pattern of knowledge and action.
Science in History 1954 (London: Watts) p 22

68   In England, more than in any other country, science is felt rather than thought .... A defect of the English is their almost complete lack of systematic thinking. Science to them consists of a number of successful raids into the unknown.
The Social Function of Science 1938, p 197

69   In fact, we will have to give up taking things for granted, even the apparently simple things. We have to learn to understand nature and not merely to observe it and endure what it imposes on us. Stupidity, from being an amiable individual defect, has become a social crime.
The Origin of Life 1967 (London: Weidenfeld & Nicholson) p 163

70   In my own field, x-ray crystallography, we used to work out the structure of minerals by various dodges which we never bothered to write down, we just used them. Then Linus Pauling came along to the laboratory, saw what we were doing and wrote out what we now call Pauling's Rules. We had all been using Pauling's Rules for about three or four years before Pauling told us what the rules were.
The Extension of Man 1972 (London: Weidenfeld & Nicolson) p 116

71   It is characteristic of science that the full explanations are often seized in their essence by the percipient scientist long in advance of any possible proof.
      The Origin of Life 1967 (London: Weidenfeld & Nicholson) p 251

72   Life is a partial, continuous, progressive, multiform and conditionally interactive, self-realisation of the potentialities of atomic electron states.
      The Origin of Life 1967 (London: Weidenfeld & Nicholson) p xv

73   Man is occupied and has been persistently occupied since his separate evolution, with three kinds of struggle: first with the massive unintelligent forces of nature, heat and cold, winds, rivers, matter and energy; secondly, with the things closer to him, animals and plants, his own body, its health and disease; and lastly, with his desires and fears, his imaginations and stupidities.
      The World, The Flesh and The Devil 1929, p 15

74   The only way of learning the method of science is the long and bitter way of personal experience.

75   The probability of the formation of a highly complex structure from its elements is increased, or the number of possible ways of doing it diminished, if the structure in question can be broken down into a finite series of successively smaller substructures.
      The Origin of Life on Earth ed A I Oparin, 1959 (Oxford: Pergamon) p 155

Claude Bernard 1813–1878

76   A living organism is nothing but a wonderful machine.
      Introduction à l'étude de la médécine expérimentale 1865, II, 1

77   *C'est la fixité du milieu interieur qui est la condition d'une vie libre et indépendante, et tous les mechanismes vitaux ... n'ont qu'un but, celui de maintenir constantes les conditions de vie dans ce milieu interne.*
      It is the constancy of the internal climate which is the condition for a free and independent life. All the mechanisms of life have only one end, that of keeping constant the conditions of life in this internal climate.
      Introduction à l'étude de la médecine expérimentale 1966 (Paris: Bordas) p 5

78   *Il y a au moins deux milieux à considerer: le milieu extérieur ou extra-organique, et le milieu intérieur ou intra-organique.*

      There are at least two *milieux* to be considered: the external or extra-organic *milieu*, and the internal or intra-organic *milieu*.
      Introduction à l'étude de la médecine expérimentale 1865, II, i, lines 3–5

79   *La science n'admet pas les exceptions; sans cela il n'y aurait aucun déterminisme dans la science, ou plutôt il n'y aurait plus de science.*
      Science allows no exceptions; without this there would be no determinism in science, or rather, there would be no science at all.
      Lecons de pathologie experimentale 1871

80   *Un poète contemporain a characterisé ce sentiment de la personalité de l'art par ces mots: l'art, c'est moi, la science, c'est nous.*
      A modern poet has characterised the personality of art and the impersonality of science as follows: Art is I; science is we.
      [Victor Hugo in *William Shakespeare* 1864]
      Introduction à l'étude de la médecine experimentalé 1865, I, 2.4, line 1742

81   Science increases our power in proportion as it lowers our pride.

Edward Bernard 17th Century

82   Books and experiments do well together, but separately they betray an imperfection, for the illiterate is anticipated unwillingly by the labours of the ancients, and the man of authors deceived by story instead of science (1671).
In S J Rigaud *Correspondence of Scientific Men of the Seventeenth Century* Oxford, 1841, vol 1, p 158

Jacques Bernouilli 1654–1705

83   *Eadem mutata resurgo.*
Though changed I shall arise the same.
[Inscribed on his tomb in the Münster, Basel, in imitation of Archimedes]
[John Goodsir copied this] [Equation of logarithmic spiral $r = a \exp(bt)$.]

84   I recognise the lion by the paw
[On seeing the anonymously published solution to his problem of the brachistochrone.]
[Newton's housekeeper: 'Bernoulli sent problem, Isaac Newton home at 4 pm, finished it by 4 am (29 January 1696)]
See *Nature* 1988 **333** 592 in Fred Hoyle *The Sciences* November 1982, p 13

85   We define the art of conjecture, or stochastic art, as the art of evaluating as exactly as possible the probabilities of things, so that in our judgments and actions we can always base ourselves on what has been found to be the best, the most appropriate, the most certain, the best advised; this is the only object of the wisdom of the philosopher and the prudence of the statesman.
Ars Conjectandi Basel, 1713. Transl in B.de Jouvenel *The Art of Conjecture* 1967 (London: Weidenfeld and Nicolson) p 21

Jons Jacob Berzelius 1779–1848

86   [On moving from collecting to experimental work] Immediately with the first participation in them I was seized with a feeling never previously experienced; I was irrevocably gripped by this method of pursuing knowledge. I must needs repeat for myself the experiments I had seen him [Ekmark] perform, and although I was unable to buy any instruments, I improvised apparatus, with his help, which I myself could make .... [When he first collected oxygen in a laboratory exercise] I ... have seldom experienced a moment of such pure and deep happiness as when the glowing stick which was thrust into it lighted up and illuminated with unaccustomed brilliancy my windowless laboratory.
Autobiographical Notes transl Olof Larsell, Baltimore, 1934, pp 7–15, 22, 25-6

Bruno Bettelheim 1903–1990

87   No longer can we be satisfied with a life where the heart has its reasons which reason cannot know. Our hearts must know the world of reason, and reason must be guided by an informed heart.
In *The Guardian* 15 March 1990, p 39

Bhartrhari 5th or 6th Century

88   In this vain fleeting universe, a man
Of wisdom has two courses: first, he can

Direct his time to pray, to save his soul,
And wallow in religion's nectar-bowl;
But, if he cannot, it is surely best
To touch and hold a lovely woman's breast,
And to caress her warm round hips, and thighs,
And to possess that which between them lies.
[D D Kosambi (1907–1966) the editor of the Sanskrit text, was an Indian mathematician of wide learning]
Transl John Brough *Poems from the Sanskrit* 1968 (Harmondsworth: Penguin) p 113

Bhaskara [Acharya] 1114–ca 1185

89   Beautiful and dear Lilavati, whose eyes are like a fawn's ...? How many are the variations of form of the god Chambhu by the exchange of his ten attributes held reciprocally in his several hands: namely the rope, the elephant hook, the serpent, the tabor, the skull, the trident, the bedstead, the dagger, the arrow and the bow ...?
Sidd'hanta-siromani chapter in *Lilavati* ca 1150, transl H T Colebrooke 1817 (London: Murray) II. 16 and XIII, 269

90   A particle of tuition conveys science to a comprehensive mind; and having reached it, expands of its own impulse. As oil poured upon water, as a secret entrusted to the vile, as alms bestowed upon the worthy, however little, so does science infused into a wise mind spread by intrinsic force.
Conclusion of *Vija-Ganita chapter* in *Lilavati* ca 1150, transl H T Colebrooke 1817 (London: Murray)

The Bible

91   ... Jews ask for signs and Greeks seek after wisdom.
I Corinthians 1:22

92   ... and should not I have pity on Nineveh, that great city; wherein are more than six score thousand persons that cannot discern between their right hand and their left hand; and also much cattle?
Jonah 4:11

93   And furthermore, my son, be admonished: of making many books there is no end; and much study is a weariness of the flesh.
Ecclesiastes 12:12

94   And I went unto the angel, saying unto him that he should give me the little book. And he saith unto me, Take it and eat it up; and it shall make thy belly bitter, but in thy mouth it shall be as sweet as honey.
Revelations 10:9

95   And one said, Rise and measure the temple of God, and the altar, and them that worship therein.
Revelations 11:1

96   And the king spake unto Ashpenaz the master of his eunuchs, that he should bring certain of the Children of Israel, and of the king's seed, and of the princes'; Children in whom was no blemish, but well favoured, and skilful in all wisdom, and cunning in knowledge, and understanding science, and such as had ability in them to stand in the king's palace, and whom they might teach the learning and the tongue of the Chaldeans.

[Those chosen were Belteshazzar, Shadrach, Meshach and Abed-nego, whose history is discussed further on in the chapter. This is the only mention of science in the Old Testament]
Daniel 1:3-4

97 And the Lord said, Behold, the people is one, and they have all one language; and this [building the Tower of Babel] they begin to do: and now nothing will be restrained from them, which they have imagined to do.
Authorised Version

98 And ye shall know the truth, and the truth shall make you free.
[Inscribed on the wall of the main lobby at the CIA Headquarters, Langley, Virginia, USA where the inhabitants say that the truth only makes them anxious.]
John 8:32

99 The chariots shall rage in the street
They shall jostle one another in the broad ways;
They shall seem like torches,
They shall run like lightnings ...
Woe to the bloody city!
It is all full of lies and robbery
... and the noise of the rattling of the wheels,
... and of the jumping chariots.
[of Nineveh]
Nahum 2:4, 3:1

100 Evil communications corrupt good manners.
[Quoted by Paul from Menander's *Thais*]
I Corinthians 15:33

101 Evil devices are an abomination to the Lord: but pleasant words are pure.
He that is greedy of gain troubleth his own house.
Proverbs 15:26–27

102 For all the Athenians and strangers which were there spent their time in nothing else, but either to tell, or to hear some new thing.
Acts 17:21

103 For he that hath, to him shall be given: and he that hath not, from him shall be taken even that which he hath.
Mark 4:25

104 For which of you, intending to build a tower, sitteth not down first, and counteth the cost, whether he have sufficient to finish it.
Luke 14:28

105 He burned the bones of the King of Edom into lime.
Amos 2:1

106 He that increaseth knowledge, increaseth sorrow.
Ecclesiastes 1:18

107 He that sinneth before his Maker, Let him fall into the hands of the physician.
Ecclesiastus (Apocrypha) 38:15

108 His breath goeth forth, he returneth to his earth; in that very day his thoughts perish.
[Psalmist on death]

109   *In principium erat verbum.*
       [Did Newton choose the title of the *Principia* from this?]
       The Vulgate John 1:1
       [In the beginning was information. Transl J Peters *Einführung in die allgemeine Informationstheorie* 1967, (Berlin: Springer) p 255]

110   *Mene, mene, tekel, upharsin.*
       [Said to be Aramaic for: numbered, numbered, weighed, divided]
       Daniel 5:25

111   O Timothy, keep that which is committed to thy trust, avoiding profane
       and vain babblings, and oppositions of science falsely so called.
       [This is the only mention of science in the New Testament—in Greek, *gnosis*]
       1 Timothy 6:20

112   O ... my desire is ... that mine adversary had written a book.
       Job 31:35

113   Out of whose womb came the ice?
       Job 38:29

114   That ye, being rooted and grounded in love, may be strong to apprehend
       with all the saints what is the breadth and length and height and depth.
       Ephesians 3:17–18 (R.V.)

115   Where there is no vision the people perish.
       Proverbs 29:18

116   Where wast thou when I laid the foundations of the earth? Declare if thou
       hast understanding. Who hath laid the measures thereof, if thou knowest?
       Or who hast stretched the line upon it? Hast thou entered into the springs
       of the sea Or hast thou walked in the search of the depth? Have the gates
       of death been opened unto thee? Or hast thou seen the doors of the shadow
       of death? Canst thou bind the sweet influences of the Pleiades, or loose the
       bands of Orion? Canst thou lift up thy voice to the clouds, that abundance
       of waters may cover thee? Canst thou send lightnings, that they may go,
       and say unto thee, Here we are? Who hath put wisdom in the inward parts?
       Or who hath given understanding to the heart? ... who can stay the bottles
       of heaven?
       Job 38:4–5, 16–17, 31, 34–37

117   Who can number the sand of the sea,
       And the drops of rain,
       And the days of eternity?
       Who can find out the height of heaven,
       And the breadth of the earth,
       and the deep, and wisdom?
       Ecclesiasticus (Apocrypha) Chapter 1

118   Who was it who measured the water of the sea in the hollow of his hand
       and calculated the dimensions of the heavens,
       gauged the whole earth to the bushel,
       weighed the mountains in scales,
       the hills in a balance?
       Isaiah 40:12 in *The Jerusalem Bible* 1966

119   For unto every one that hath shall be given, and he shall have abundance:
       but from him that hath not shall be taken away even that which he hath.

[Matthew Principle of scientific publication enunciated by R K Merton]
Matthew 25:29

120   But by measure and number and weight thou didst order all things.
Wisdom of Solomon (Apocrypha) 11:20

Al-Biruni 973–1048

121   Once a sage was asked why scholars always flock to the doors of the rich,
whilst the rich are not inclined to call at the doors of scholars. 'The scholars'
he answered, 'are well aware of the use of money, but the rich are ignorant
of the nobility of science'.

122   [On the science and culture of the Hindus] I can only compare their astro-
nomical and mathematical literature ... to a mixture of pearl shells and
sour dates, or of costly crystals and common pebbles. Both kinds of things
are equal in their eyes, since they cannot rise themselves to the methods of
strictly scientific deduction.
Hindustan transl C E Sachau, London, 1888

Otto von Bismarck 1815–1893

123   *Die Politik ist keine exakte Wissenschaft.*
Politics is not an exact science.
Speech, Prussian Chamber, 10 December 1803

Patrick Maynard Stuart Blackett 1897–1974

124   A first-rate laboratory is one in which mediocre scientists can produce
outstanding work.
Quoted by M G K Menon in his commemoration lecture on H J Bhabha, Royal Institution, 1067

125   The scientist can encourage numerical thinking on operational matters and
so help avoid running the war by gusts of emotion.
Operational Research in the RAF (London: HMSO)

Blackstone [The Magician]

126   The fundamental principle of the magician's art is misdirection.

William Blake 1751–1827

127   [On industrialisation] Hampstead, Highgate, Finchley, Muswell Hill rage
loud
Before Bromion's iron Tongs and glowing Poker reddening fierce; ...
Jerusalem plate 16, 1.1–2

128   The Atoms of Democritus
And Newton's Particles of Light
Are sands upon the Red Sea shore
Where Israels's tents do shine so bright.
Mock on, Mock on: Voltaire, Rousseau ca 1800

129   Consider Sexual Organization and hide thee in the dust.
       Jerusalem plate 34, Chapter 2

130   He who would do good to another must do it in Minute Particulars:
       General Good is the plea of the scoundrel, hypocrite and flatterer,
       For Art and Science cannot exist but in minutely organised particulars.
       Jerusalem plate 55, 60–61

131   I must Create a System, or be enslaved by another Man's;
       I will not Reason and Compare; My business is to Create.
       Jerusalem plate 10, 20

132   I was in a Printing-house in Hell, and saw the method in which knowledge
       is transmitted from generation to generation.
       The Marriage of Heaven and Hell

133   Science is divided into Bowlahoola and Allamand.
       Milton

134   To teach doubt and Experiment
       Certainly was not what Christ meant.
       Blake Complete Writings Nonesuch edn, p 752

135   Who publishes doubt and calls it knowledge whose Science is Despair.
       Milton II

136   [Beulah] ... a place where contrarieties are equally true.
       Milton II, plate 30

137   What is now proved was once only imagined.
       The Marriage of Heaven and Hell

Hendrik Wade Bode, Frederick Mosteller, John Tukey and Charles Winsor

138   Recapture the universalist spirit of the early natural philosophers.
       Learn *science* and not *sciences*.
       Know in capsule form the dozen central concepts of each of the major
       sciences.
       Learn the habits of mind of the chemist, psychologist and geologist.
       Use in each science some of the intellectual equipment of the other sciences.
       Be exceptional in breadth of appreciation.
       Be able in biological and medical science to suggest physical explanations
       or mathematical models for known or conjectured facts.
       Be familiar with forging and milling, the functions of a turret lathe, the
       kinds of heat treating used and their effects, what an industrial still looks
       like ...
       The education of a scientific generalist in *Science* 1949 **109** 553–8

Aleksandr Aleksandrovich Bogdanov [Malinovskii] 1873–1928

139   Nature is a unity—of great and small, of living and dead.
       Tektologia vseobshchaya organizatsionnaya nauka (last words) 1922 2nd edn (Moscow:
       Grzhebina) p 572

David Joseph Bohm 1917–

140   There are no things, only processes.
       In C H Waddington *The Evolution of an Evolutionist* 1975 (Edinburgh: Edinburgh
       University Press) p 4

Niels Henrik David Bohr 1885–1962

141   ... two sorts of truth: trivialities, where opposites are obviously absurd,
       and profound truths, recognised by the fact that the opposite is also a
       profound truth.
       [Also Thomas Mann]
       In Hans Bohr *My Father* in *Niels Bohr: his life and work* ... ed S Rozental, 1967
       (New York: Wiley) p 328

142   An expert is a man who has made all the mistakes, which can be made, in
       a very narrow field.
       Edward Teller, 10 October 1972, US Embassy

143   When it comes to atoms, language can be used only as in poetry. The
       poet too, is not nearly so concerned with describing facts as with creating
       images.
       In J Bronowski *The Ascent of Man* 1975 (London: BBC) p 340

144   Every sentence I utter must be understood not as an affirmation but as a
       question.

Ludwig Boltzmann 1844–1906

145    $S = k \log \Omega$.
       [Carved above his name on his tombstone in the Zentralfriedhof in Vienna]

Wolfgang Bólyai 1775–1856

146   Detest it just as much as lewd intercourse; it can deprive you of all your
       leisure, your health, your rest, and the whole happiness of your life.
       [To his son János, warning him to give up his attempts to prove the Euclidean postulate
       on parallels]

Napoleon Bonaparte 1769–1821

147   The advance and perfecting of mathematics are closely joined to the pros-
       perity of the nation.

148   [To Laplace, on receiving a copy of the *Mécanique Céleste*] The first six
       months which I can spare will be employed in reading it.
       In M Crosland *The Society of Arcueil* 1967 (Cambridge: Cambridge University Press)

[Sir] Hermann Bondi 1919–

149   [Science doesn't deal with facts; indeed] fact is an emotion-loaded word
       for which there is little place in scientific debate. Science is above all a
       cooperative enterprise.
       Nature 1977 **265** 286

John Tyler Bonner 1920–

150   We've got to have some sort of morality soon that says that people who
       have more than two children are beasts. It has to come.

Étienne Bonnot [Abbé de Condillac] 1714–1780

151   *Une science bien traitée n'est qu'une langue bienfaite.*
A science well-expounded is just like a well-made language.
Traité des Systèmes XVIII

152   *Voulez vous apprendre les sciences avec facilité? Commencez par apprendre votre langue.*
Do you wish to learn science easily? Then begin by learning your own language.
Essai sur l'origine des connaissances humaines

Andrew Donald Booth 1918–

153   Every system is its own best analogue.

Jorge Luis Borges 1899–1986

154   Every writer creates his own precursors.
Labyrinths 1970 (Harmondsworth: Penguin) p 10

155   *Mis libros (que no saben que yo existo)*
son tan parte de mi como este rostro
de sienes grises y de grises ojos
que vanamente busco en los cristales.
My books (which do not know that I exist) are as much a part of me as is this face, the temples gone grey and the eyes grey, the face I vainly look for in the mirror . . . .
Mis Libros, The Book of Sand 1979 (Harmondsworth: Penguin) p 178

Max Born 1882–1970

156   I am now convinced that theoretical physics is actual philosophy.
Autobiography

Roger Joseph Boscovich 1711–1780

157   *Homo hominem arreptum a Tellure, et ubicumque exigua impulsum vi vel une etiam oris flatu impetitum, ab hominum omnium commercio in infinitum expelleret, nunquam per totam aeternitatem rediturum.*
Were it not for gravity one man might hurl another by a puff of his breath into the depths of space, beyond recall for all eternity.
Theoria par 552

158   It will be found that everything depends on the composition of the forces with which the particles of matter act upon one another; and from these forces, as a matter of fact, all phenomena of Nature take their origin.
Theoria Philosophiae Naturalis Venice, 1763, sec. I.5

James Boswell 1740–1795

159   When Dr Johnson felt, or fancied he felt, his fancy disordered, his constant recurrence was to the study of arithmetic.
Life of Johnson Harper's edn, 1871, vol 2, p 264

Gordon Bottomley 1874–1949

160   Your worship is your furnaces,
Which, like old idols, lost obscenes,

Have molten bowels; your vision is
Machines for making more machines.
To Iron Founders and Others in *Poems of Thirty Years* 1925 (London: Constable)

[Sir] Thomas Bouch 1822–1880

161 [Designer of the Tay Bridge which collapsed in 1879, was asked why he used
that particular design. He brought great derision on himself by replying that
it was] because it facilitated the calculations.
In *The Guardian* (letter) 13 February 1988

Kenneth R Boulding 1910–

162 [To a student who complained about his examination grading] Well, this is
an unjust world and education is intended to prepare you for it.
Technology Review October 1979

Pierre Boulez 1925–

163 Music cannot move forward without science.
The Observer 27 July 1975, p 17

Matthew Boulton 1711–1780

164 I am selling what the whole world wants; power.
[Letter to Catherine the Great of Russia offering steam engines for sale]
In J Bernal *Science in History* (Cambridge, MA: MIT Press)

Nicholas Bourbaki (pseudonym)

165 Structures are the weapons of the mathematician.
[Collective pseudonym of the Nancy school of mathematics. See *Scientific American*
May 1957, p 88]

Cecil Maurice Bowra 1898–1971

166 [Used to say that] scientists were treacherous allies on committees, for they
were apt to change their minds in response to arguments.

Robert Boyle 1627–1691

167 Father of Chemistry and Uncle of the Earl of Cork.
[On his tombstone]

Francis Herbert Bradley 1846–1924

168 Metaphysics is the finding of bad reasons for what we believe upon instinct.
Appearance and Reality Preface

[Sir] William Henry Bragg 1862–1942

169 Physicists use the wave theory on Mondays, Wednesdays and Fridays, and
the particle theory on Tuesdays, Thursdays and Saturdays.

170 After a year's research, one realises that it could have been done in a week.
In *Fifty Years of X-ray Diffraction* ed P P Ewald, Utrecht, 1962

[Sir] William Lawrence Bragg 1890–1971

171 The electron is not as simple as it looks.
[Recounted by Sir George Paget Thompson at electron diffraction conference, 1967]

172   The important thing in science is not so much to obtain new facts as to
        discover new ways of thinking about them.
        In A Koestler and J R Smithies *Beyond Reductionism* 1958 (London: Hutchinson) p
        115

173   [To his sister] You and I find things easier than people.
        In B M Caroe *William Henry Bragg* 1978 (Cambridge: Cambridge University Press)
        p 39

174   The dividing line between the wave or particle nature of matter and radia-
        tion is the moment 'Now'. As this moment steadily advances through time
        it coagulates a wavy future into a particle past.
        The Development of X-ray Analysis 1975 (London: Bell)

175   The biggest mistake of my scientific career.
        [To make his model of the α-helix with an exact four-fold axis instead of a non-integer
        screw.]
        In Francis Crick *What Mad Pursuit* 1988 (London: Weidenfeld & Nicolson) p 58

176   I sometimes feel it necessary to remind young research students that we
        are not writing our papers for consideration only by God and a committee
        of archangels.

Tycho Brahe 1546–1601

177   And when statesmen or others worry him [the scientist] too much, then
        he should leave with his possessions. With a firm and steadfast mind one
        should hold under all conditions, that everywhere the earth is below and
        the sky above, and to the energetic man, every region is his fatherland.
        [The 'brain drain' has existed as long as science]
        Denmark, 1597

178   Now it is quite clear to me that there are no solid spheres in the heavens,
        and those that have been devised by the authors to save the appearances,
        exist only in the imagination, for the purpose of permitting the mind to
        conceive the motion which the heavenly bodies trace in their courses.

Georges Braque 1882–1963

179   *L'art est fait pour troubler. La science rassure.*
        Art upsets, science reassures.
        Pensées sur l'Art (Paris: Gallimard)

180   *La vérité existe. On n'invente que la mensonge.*
        The truth exists—only fictions are invented.
        Pensées sur l'Art (Paris: Gallimard)

Bertold Brecht 1898–1956

181   Out of the libraries strike the slaughterers.
        Mothers gaze numbly at the skies
        For the inventions of the scholars.

182   *Und was nützt freie Forschung ohne freie Zeit zu forschen?*
        What good is freedom to research without free time to do it in?
        Leben des Galilei 1956, scene 1

183   *Es kommen die Herren Gelehrten*
        Mit falschen Teutonenbärten

und furchterfülltem Blick
Sie wollen nicht eine richtige
Sondern eine arisch gesichtige
Genehmigte deutsche Physik.
The scholars come with false teutonic beards and eyes
filled with fear. They are not interested in real science but
in an Aryan, authorised German physics.
Furcht und Elend des III Reiches (Berlin: Suhrkamp)

Sydney Brenner 1927–

184   Progress in science depends on new techniques, new discoveries and new
       ideas, probably in that order.
       In *Nature* 5 May 1980 and 7 June 1990

185   There will be no difficulty in computers being adapted to biology. There
       will be Luddites. But they will be buried.
       In H Judson *The Eighth Day of Creation* p 221

186   Have you tried neuroxing papers? It's a very easy and cheap process. You
       hold the page in front of your eyes and you let it go through there into the
       brain. It's much better than xeroxing.
       In L Wolpert and A Richards (eds) *A Passion for Science* 1988 (Oxford: Oxford
       University Press) p 104

[Sir] David Brewster 1781–1868

187   And why does England thus persecute the votaries of her science? Why
       does she depress them to the level of her hewers of wood and her drawers
       of water? It is because science flatters no courtier, mingles in no political
       strife? ... Can we behold unmoved the science of England, the vital prin-
       ciple of her arts, struggling for existence, the meek and unarmed victim of
       political strife?
       Quarterly Review 1830 **43** 320, 323–4, 341 (reviewing Babbage's book *Reflexions on
       the Decline of Science in England*

188   The infant [Newton] ... ushered into the world was of such diminutive size,
       that, as his mother afterwards expressed it to Newton himself, he might
       have been put into a quart-mug ....
       Memoirs of Newton 1855

189   There is no profession so incompatible with original enquiry as is a Scotch
       Professorship, where one's income depends on the numbers of pupils. Is
       there one Professor in Edinburgh pursuing science with zeal? Are they not
       all occupied as showmen whose principal object is to attract pupils and
       make money?
       In *Physics Bulletin* 1983 **34** 511

Leonid Ilich Brezhnev 1906–1982

190   There is nothing more practical than a good theory.
       In V Rich *Nature* 1977 **270** 470–1

Gérard Bricogne

191   Mankind is a catalysing enzyme for the transition from a carbon-based to
       a silicon-based intelligence.

Robert Bridges 1844–1930

192 ... we only think to find
A structure of blind atoms to their habits enslaved.
Testament of Beauty 1930

193 Now will the Orientals make hither in return
Outlandish pilgrimage; their wise acres have seen
The electric light; in the West, and come to worship.
The Testament of Beauty 1930, line 592

Anthelme Brillat-Savarin 1755–1826

194 *La destinée des nations dépend de la manière dont elle se nourissent.*
The destiny of countries depends on the way they feed themselves.
Physiologie du Goût 1825

195 *La découverte d'un mets nouveau fait plus pour le bonheur du genre humain
que la découverte d'une étoile.*
The discovery of a new dish does more for human happiness than the
discovery of a star.
Physiologie du Goût 1825, IX

Léon Brillouin 1889–1969

196 *Un système vivant est un système ouvert et pourtant stable. On peut le
comparer à une flamme.*
A living system is an open yet stable system. It could be likened to a flame.
Vie, matière et observation (Paris: Albin Michel) Chapter 3, 7

Louis Victor de Broglie 1892–1987

197 Two seemingly incompatible conceptions can each represent an aspect of
the truth .... They may serve in turn to represent the facts without ever
entering into direct conflict.
Dialectica I, 326

Jacob Bronowski 1908–1974

198 ... no science is immune to the infection of politics and the corruption
of power .... The time has come to consider how we might bring about a
separation, as complete as possible, between Science and Government in all
countries. I call this the disestablishment of science, in the same sense in
which the churches have been disestablished and have become independent
of the state.
Encounter July 1971, p 15

199 By the worldly standards of public life, all scholars in their work are of
course oddly virtuous. They do not make wild claims, they do not cheat,
they do not try to persuade at any cost, they appeal neither to prejudice
nor to authority, they are often frank about their ignorance, their disputes
are fairly decorous, they do not confuse what is being argued with race,
politics, sex or age, they listen patiently to the young and to the old who
both know everything. These are the general virtues of scholarship, and
they are peculiarly the virtues of science.
Science and Human Values 1961 (London: Hutchinson) p 67

200 The hand is the cutting edge of the mind.
The Ascent of Man 1975 (London: BBC) p 116

201 It is important that students bring a certain ragamuffin, barefoot irreverence to their studies; they are not here to worship what is known, but to question it.
The Ascent of Man 1975 (London: BBC) p 360

202 Man masters nature not by force but by understanding. That is why science has succeeded where magic failed: because it has looked for no spell to cast on nature.
Science and Human Values 1956 (London: Hutchinson) p 20

203 Science has nothing to be ashamed of, even in the ruins of Nagasaki.
Science and Human Values 1959 (New York: Harper and Row) p 93

204 To me, being an intellectual doesn't mean knowing about intellectual issues; it means taking pleasure in them.
Magic, Science and Civilisation 1978 (New York: Columbia University Press) p 36

Jacob Bronowski and Bruce Mazlish 1908–1974 and 1923–

205 Every thoughtful man who hopes for the creation of a contemporary culture knows that this hinges on one central problem: to find a coherent relation between science and the humanities.
The Western Intellectual Tradition 1960 (London: Hutchinson) p vii

Rupert Brooke 1887–1915

206 But somewhere, beyond Space and Time
Is wetter water, slimier slime.
And there (they trust) there swimmeth One
Who swam ere rivers were begun,
Immense, of fishy form and mind,
Squamous, omnipotent, and kind.
Heaven

207 For Cambridge people rarely smile,
Being urban, squat, and packed with guile;
The Old Vicarage, Grantchester written Berlin, 1912

George Spencer Brown 1923–

208 To arrive at the simplest truth, as Newton knew and practised, requires *years of contemplation.* Not activity. Not reasoning. Not calculating. Not busy behaviour of any kind. Not reading. Not talking. Not making an effort. Not thinking. Simply *bearing in mind* what it is one needs to know. And yet those with the courage to tread this path to real discovery are not only offered practically no guidance on how to do so, they are actively discouraged and have to set about it in secret, pretending meanwhile to be diligently engaged in the frantic diversions and to conform with the deadening personal opinions which are continually being thrust upon them.
The Laws of Form 1969 (London: Allen & Unwin)

[Sir] Thomas Browne 1605–1682

209 ... indeed what reason may not go to Schoole to the wisdome of Bees, Aunts and Spiders? what wise hand teacheth them to doe what reason

cannot teach us? ruder heads stand amazed at those prodigious pieces of
nature, Whales, Elephants, Dromidaries and Camels; these I confesse, are
the Colossus and Majestick pieces of her hand; but in these narrow Engines
there is more Mathematicks, and the civilitie of these little Citizens more
neatly sets forth the wisedome of their Maker.
[Browne's writings are full of curious pre- and proto-scientific learning]
Religio Medici I, 15

210   God is like a skilful Geometrician.
[cf Plutarch *Symposiaes* viii, 2: How Plato is to be understood, when he saith: That
God continually is exercised in Geometry. It is not, however, in Plato's works]
Religio Medici I, 16

211   All things began in Order, so shall they end, and so shall they begin again,
according to the Ordainer of Order, and the mystical mathematicks of the
City of Heaven.
Hydriotaphia, Urn-burial and the Garden of Cyrus 1896 (London: Macmillan) p 160

212   Sure there is music even in the beauty, and the silent note which Cupid
strikes, far sweeter than the sound of an instrument. For there is music
wherever there is harmony, order and proportion; and thus far we may
maintain the music of the spheres; for those well ordered motions, and regular paces, though they give no sound unto the ear, yet to the understanding
they strike a note most full of harmony.
Religio Medici II, 9

213   Thus is Man that great and true Amphibian whose nature is disposed to
live ... in divided and distinguished worlds.
Religio Medici I, 34

214   What song the Syrens sang, or what name Achilles assumed when he hid
himself among women, though puzzling questions, are not beyond all con-
jecture.
[Asked first by Tiberius. Suetonius *Tiberius* LXX. For proposed answers see Robert
Graves *The White Goddess* 1948]
Hydriotaphia, Urn-burial and the Garden of Cyrus 1658, Chapter 5

Robert Browning 1812–1889

215   Ignorance is not ignorance but sin.
The Inn Album V

216   Some lump, ah God, of lapis lazuli
Big as a Jew's head cut off at the nape,
Blue as a vein o'er the Madonna's breast ...

Giordano Bruno 1548–1600

217   *Se no è vero ma è ben trovato.*
It may not be true but it is well done.
[Attributed]

218   And just as in the case of letters, it is not necessary that there be many
kinds and shapes of minims in order to form innumerable species.
De Minimo in P Redondi *Galileo, Heretic* p 60

Janina Brzostowska 1907–

219   Patina,
       August,
       mellow green,
       flowering on church spires
       and on rooftops
       billowing through the centuries.
       Necklace, virescent
       as eucalyptus leaves,
       patiently burnished,
       lustrous,
       warm
       on the skin of a woman.
       And lit by the desert sun
       gray-green columns
       of eastern temples
       recalling the ancient Persians.
       [Malachite—basic copper carbonate]
       Malachite transl from the Polish by Neil S Snider in W Brostow *Structure of Materials*
       1979 (New York: Wiley) p 164

John Buchan [Lord Tweedsmuir] 1875–1940

220   To live for a time close to great minds is the best kind of education.
       Memory Hold the Door 1940, p 34

Robert Buchanan 1841–1901

221   Alone at nights, I read my Bible more and Euclid less.
       An Old Dominie's Story

Ludwig Büchner 1824–1899

222   *Ohne Phosphor, Kein Gedanke.*
       Without phosphorus there would be no thoughts.
       [Attributed]

Art Buchwald 1925–

223   War is too serious a business to be left to computers.
       International Herald Tribune 14/15 June 1980

Buddha ca 563–483 BC

224   All composite things decay. Strive diligently.
       [His last words]

Georges Leclerc [Comte de] Buffon 1707–1788

225   ... all the work of the crystallographers serves only to demonstrate that
       there is only variety everywhere where they suppose uniformity ... that in
       nature there is nothing absolute, nothing perfectly regular.
       Histoire Naturelle des Minéraux Paris, 1783–88, III, 433

226   One can descend by imperceptible degree from the most perfect creature to the most shapeless matter, from the best-organised animal to the roughest mineral.
De la Manière d'étudier et de Traiter l'Histoire Naturelle in *Oeuvres Complètes* Paris, 1774–9, I, p 17

227   *Les ouvrages bien écrits seront seuls qui passeront à la posterité: la quan-tité des connaissances, la singularité des faits, la nouveauté même des découvertes ne sont pas de surs garants de l'immortalité. Ces choses sont hors de l'homme, le style c'est l'homme même.*
Only those works which are well-written will pass to posterity: the amount of knowledge, the uniqueness of the facts, even the novelty of the discov-eries are no guarantees of immortality. These things are exterior to a man but style is the man himself.
Discours sur le style 1753

Nikolai Ivanovich Bukharin 1888–1938

228   ... we must do our utmost to promote the union of science with technique and with the organisation of production. *Communism signifies intelligent, purposive and consequently, scientific production. We shall, therefore, do everything in our power to solve the problem of the scientific organisation of production.*
Azbuka Kommunizma 1920 Peterburg, sec. 102, ed E H Carr *ABC of Communism* 1969 (Harmondsworth: Penguin) p 347

[Sir] Edward Bullard 1907–1980

229   I should be prepared to argue that Rutherford was a disaster. He started the 'something for nothing' tradition which I was brought up in and had some difficulty freeing myself from—the notion that research can always be done on the cheap, the notion that things don't cost what they do cost. It is wrong. The war taught us differently. If you want quick and effective results you must put the money in.
In P Grosvenor and J McMillan *The British Genius* 1973 (London: Dent)

Edward Bulwer-Lytton 1803-1873

230   In science, read, by preference, the newest works; in literature, the oldest.
Caxtoniana Essay X

John Bunyan 1628–1688

231   Some said, 'John, print it', others said, 'Not so'; Some said, 'It might do good', others said, 'no'.
Apology for his Book

Luis Buñuel 1900–1989

232   I find it [science] analytic, pretentious, superficial, largely because it does not address itself to dreams, chance, laughter, feelings or paradox—all the things I love the most.
Last Sigh autobiography

Luther Burbank 1849–1926

233   Heredity is just environment stored.

Edmund Burke 1729–1797

234 The age of chivalry is gone. That of sophisters, economists and calculators
has succeeded: and the glory of Europe is extinguished for ever.
Reflections on the Revolution in France 1970 (London: Dent)

235 In the groves of *their* academy, at the end of every walk, you see nothing
but the gallows.
Reflections on the Revolution in France 1970 (London: Dent)

236 A woman is but an animal, and an animal not of the highest order.
Reflections on the Revolution in France 1970 (London: Dent)

[Sir] Frank Macfarlane Burnet 1899–

237 There is virtually nothing that has come from molecular biology that can
be of any value to human living in the conventional sense of what is good,
and quite tremendous possibilities of evil, again in the conventional sense.
Changing Patterns 1968 (London: Heinemann) p 176

Daniel Burnham 1846–1912

238 Make no little plans, they have no power to stir men's souls.
[Deviser of lake front park in Chicago]
In Charles Moore *Daniel H Burnham* 1921 (Boston, MA: Houghton Muffin)

Robert Burns 1759–1796

239 Facts are chiels that winna ding, an' downa be disputed.
[Facts are entities which cannot be manipulated or disputed.]
A Dream 30

240 I waive the quantum o' the sin
The hazard of concealing
But oh: it hardens a' within
And petrifies the feeling.
Epitaph to a Young Friend 6

241 Some books are lies frae end to end,
An' some great lies were never penn'd:
Ev'n ministers they ha'e been kenn'd,
In holy rapture,
A rousing whid at times to vend,
An' nail't wi' Scripture.
[whid = lie]
Death and Doctor Hornbook 1785

Vannevar Bush 1890–1974

242 The greatest event in the world today is not the awakening of Asia, nor
the rise of communism—vast and portentious as those events are. It is the
advent of a new way of living, due to science, a change in the conditions of
work and the structure of society which began not so very long ago in the
West, and is now reaching out over all mankind.

243 We have no national policy for science.
Science—the Endless Frontier. A report to the President July 1945 (Washington,
DC: US Govt Printing Office) p 7

244   Knowledge for the sake of understanding, not merely to prevail, that is
      the essence of our being. None can define its limits, or set its ultimate
      boundaries.
      Science Is Not Enough 1967

Samuel Butler 1612–1680

245   In mathematics he was greater
      Than Tycho Brahe, or Erra Pater:
      For he, by geometric scale,
      Could take the size of pots of ale;
      Resolve, by sines and tangents straight,
      If bread or butter wanted weight;
      And wisely tell what hour o' the day
      The clock does strike, by Algebra.
      Hudibras 1663, part 1

246   A learned society of late,
      The glory of a foreign state,
      Agreed, upon a summer's night,
      To search the Moon by her own light.
      The Elephant in the Moon ca 1676

Samuel Butler 1835–1902

247   A hen is only an egg's way of making another egg.
      Life and habit VIII

248   We shall never get people whose time is money to take much interest in
      atoms.

249   X-rays. Their moral is this—that a right way of looking at things will see
      through almost anything.
      Notebooks vol V

Herbert Butterfield 1900–

250   It [the Scientific Revolution] outshines everything since the rise of Chris-
      tianity and reduces the Renaissance and the Reformation to the rank of
      mere episodes, mere internal displacements, within the system of medieval
      Christendom .... It looms so large as the real origin of the modern world
      and of the modern mentality that our customary periodisation of European
      history has become an anachronism and an encumbrance.
      The Origins of Modern Science 1949 (London: Bell)

George Gordon [Lord] Byron 1788–1824

251   'Tis a pity learned virgins ever wed
      With persons of no sort of education,
      Or gentlemen, who, though well born and bred,
      Grow tired of scientific conversation.
      Don Juan I, XXII

252   'Tis pleasant, sure, to see one's name in print;
      A book's a book, although there's nothing in't.
      English Bards and Scotch Reviewers line 51

253   I stood in Venice, on the 'Bridge of Sighs';
A palace and a prison on each hand;
I saw from out of the wave her structures rise
As from the stroke of the Enchanter's wand.
Childe Harold's Pilgrimage opening of Canto 4

254   That soft bastard Latin
Which melts like kisses from a female mouth.
[The Italian language]
Beppo verse 44

255   When Newton saw an apple fall, he found ...
A mode of proving that the earth turn'd round
In a most natural whirl, called gravitation,
And thus is the sole mortal who could grapple
Since Adam, with a fall or with an apple.
Don Juan 10, 11

256   I suppose we shall soon travel by air-vessels; make air- instead of sea-voyages; and at length find our way to the moon; in spite of the want of atmosphere.
1822, quoted at Washington Aerospace Museum

257   Knowledge is not happiness, and science
But an exchange of ignorance for that
Which is another kind of ignorance.
Manfred

258   [On the Fellows of Trinity College, Cambridge]
The sons of science, those who thus repaid,
Linger in ease in Granta's sluggish shade;
Where on Cam's sedgy banks supine they lie,
Unknown, unhonoured live, unwept-for die;
Dull as the pictures, which adorn their halls.
Thoughts suggested by a College Examination

Alexander Graham Cairns-Smith 1931–

1   It is proposed that life on Earth evolved through natural selection from inorganic crystals.
    J. Theor. Biol. 1966 **10** 53–88

Nigel David Ritchie Calder 1931–

2   All in all, the refurbished creation myth owes more to Groucho than to Karl Marx. It is a tale of hungry molecules making dinosaurs and remodelling them as ducks; also of cowboys who put to sea, quelled the world with a magnetic needle, and then wagered their genes against a mushroom cloud the knowledge was a Good Thing.
    Timescale: An Atlas of the Fourth Dimension 1983

George John Douglas Campbell [8th Duke of Argyll] 1823–1900

3   [*Survival of the fittest*—Herbert Spencer's coinage] Nothing could be happier than this invention for ... giving vogue to whatever it might be supposed to mean .... It is the fittest of all phrases to survive.
    Organic evolution cross-examined 1898

Thomas Campbell 1777–1844

4   O Star-eyed Science! hast thou wandered there,
    To waft us home the message of despair?
    Pleasures of Hope pt 2, line 325

Albert Camus 1913–1960

5   An intellectual is someone whose mind watches itself.
    Notebooks 1935–42 transl P Thoday, 1966 (London: Hamish Hamilton) I, p 28

Karel Čapek 1860–1927

6   Rossum's Universal Robots.
    [The origin of the word 'robot']
    R.U.R. 1920 (Oxford: Oxford University Press)

7   Countless laws of construction and constitution penetrate matter like secret flashes of mathematical lightning. To equal nature it is necessary to be mathematically and geometrically exact. Number and fantasy, law and quantity, these are the living creative strengths of nature; not to sit under a green tree but to create crystals and to form ideas, that is what it means to be at one with nature.

Girolamo Cardano 1501–1576

8   To throw in a fair game at Hazards only three-spots, when something great is at stake, or some business is the hazard, is a natural occurrence and deserves to be so deemed; and even when they come up the same way for a second time if the throw be repeated. If the third and fourth plays are the same, surely there is occasion for suspicion on the part of a prudent man.
    [Who 'was ever hot-tempered, single-minded and given to women']
    De Vita Propria Liber

Henry Carey died 1743

9   Aldeborontiphoscophornio; Where left you Chrononhotonthologos?
     [Author of *God Save the King*]
     Chrononhotonthologos 1734, sc 1

Rémy Louis Carle 1930–

10   [A director of Électricité de France, asked whether the public were consulted
     on the siting of a nuclear power station]
     You don't ask the frogs when you drain the marsh.
     The Guardian 1986

Thomas Carlyle 1795–1881

11   Genius ... means transcendent capacity of taking trouble.
     Life of Frederick the Great Chapter 3

12   In a symbol there is concealment and yet revelation: here therefore, by
     Silence and by Speech acting together, comes a double significance.
     Sartor Resartus III, iii

13   It is a mathematical fact that the casting of this pebble from my hand
     alters the centre of gravity of the universe.
     [Does it?]
     Sartor Resartus III

14   The Social Science, not a 'gay science' ...; no, a dreary, desolate, and indeed
     quite abject and distressing one; what we might call, by way of eminence,
     the dismal science.
     Miscellanies, The Nigger Question

15   Such I hold to be the genuine use of gunpowder; that it makes all men alike
     tall.

16   Man is a tool using animal .... Without tools he is nothing, with tools he
     is all.
     Sartor Resartus I, Chapter 5

17   'Thought, he [Dr. Cabanis] is inclined to hold, is still secreted by the brain;
     but the Poetry and Religion (and it is really worth knowing) are a product
     of the smaller intestines.'
     Signs of the Times

Alexis Carrel 1873–1944

18   *L'éminence même d'un spécialiste le rend plus dangereux.*
     The mere eminence of a specialist makes him the more dangerous.
     L'homme cet inconnu (Paris: Plon) p 6

19   *Le meilleur moyen d'augmenter l'intélligence des savants serait de diminuer
     leur nombre.*
     The best way of increasing the [average] intelligence of scientists would be
     to reduce their number.
     L'homme, cet inconnu (Paris: Plon) Chapter 2, 4

20   *L'intélligence est presque inutile a celui qui ne possède qu'elle.*
     Intelligence is almost useless to the person whose only quality it is.
     L'homme, cet inconnu (Paris: Plon) Chapter 4, 6

Lewis Carroll [Charles Lutwidge Dodgson] 1832–1898

21   What I tell you three times is true.
       The Hunting of the Snark

22   'Can you do addition ?' the White Queen asked. 'What's one and one and
       one and one and one and one and one and one and one and one ?' 'I don't
       know', said Alice 'I lost count'.
       Through the Looking Glass p 232

23   'Why', said the Dodo, 'the best way to explain it is to do it.'
       Alice in Wonderland p 33

24   'Would you tell me, please, which way I ought to go from here ?' 'That
       depends a good deal on where you want to get to', said the Cat. 'I don't
       much care where . . . ', said Alice. 'Then it doesnt matter which way you go',
       said the Cat. 'So long as I get somewhere', Alice added as an explanation.
       'Oh, you're sure to do that', said the Cat, 'If you only walk long enough.'
       Alice's Adventures in Wonderland p 64

25   'It's too late to correct it,' said the Red Queen; 'when you've said a thing,
       that fixes it, and you must take the consequences.'
       'Complete Works of Lewis Carroll 1939 (London: Nonesuch) p 234

26   . . . It next will be right
       To describe each particular batch:
       Distinguishing those that have feathers, and bite,
       From those that have whiskers, and scratch.
       The Hunting of the Snark

Rachel Louise Carson 1907–1964

27   Silent Spring.
       [Book title]
       Silent Spring 1962 (Boston, MA: Houghton Miffin)

Charles Frederick Carter 1919–

28   In use of equipment especially, insufficient attention is paid to real costs. In
       some cases it would be cheaper not to install the equipment, but whenever
       it is needed to send each student in a separate chauffeur-driven Rolls-Royce
       to use the same equipment at another institution.
       Universities and Productivity University Conference, Committee of Vice-Chancellors
       and the Association of University Teachers, 21 March 1968

Cato [the Censor] 234–149 BC

29   How could one [Roman] haruspex look another in the face without laugh-
       ing?
       [A haruspex divined the future from the entrails of animals]
       Ascribed by Cicero *De Diviniatione* ii, 24

Raymond Bernard Cattell 1905–

30   But psychology is a more tricky field, in which even outstanding authorities
       have been known to run in circles, describing things which everyone knows
       in language which no one understands.
       The Scientific Analysis of Personality 1965 (Harmondsworth: Pelican) p 18

Christopher Caudwell [St. John Sprigg] 1908–1937

31   to the bourgeois ... art and science appear not as creative opposites but
as eternal antagonists
Studies in a Dying Culture 1938 (London: Bodley Head)

John J Cavanaugh

32   Even casual observation of the daily newspapers and the weekly news mag-
azines, leads a Catholic to ask, where are the Catholic Salks, Oppenheimers,
Einsteins?
Time 30 December 1957, p 50

William Caxton ca 1422–1491

33   The Dictes or sayengis of the philosophres
[The title of the first dated book printed in England]
1477

Louis-Ferdinand Céline 1894–1961

34   *Entre le pénis et les mathématiques ... il n'existe rien. Rien! C'est le vide.*
Voyage au bout de la nuit (Paris: Gallimard)

Miguel de Cervantes 1547–1616

35   Let us come now to references to authors, which other books contain and
yours lacks. The remedy for this is very simple; for you have nothing else to
do but look for a book which quotes them all from A to Z, as you say. Then
you put this same alphabet into yours .... And if it serves no other purpose,
at least that long catalogue of authors will be useful to lend authority to
your book at the outset.
Don Quixote transl J M Cohen (Harmondsworth: Penguin) Prologue

Paul Cézanne 1839–1906

36   ... *Traiter la nature par le cylindre, la sphère, le cône, le tout mis en perspective .... La nature, pour nous hommes, et plus en profondeur qu'en surface.*
Treat nature in terms of the cylinder, the sphere, the cone, all in perspective .... Nature, for us, lies more in depth than on the surface.
In Emile Bernard *Paul Cézanne* 1925

George Phillip [Air Vice-Marshal] Chamberlain 1905–

37   *Boffin:* A Puffin, a bird with a mournful cry, got crossed with a Baffin, a mercifully obsolete Fleet Air Arm aircraft. Their offspring was a Boffin, a bird of astonishingly queer appearance, bursting with weird and sometimes inopportune ideas, but possessed of staggering inventiveness, analytical powers and persistance. Its ideas, like its eggs, were conical and unbreakable. You push the unwanted ones away, and they just roll back.
In R W Clark *The Rise of the Boffins* 1962 (London: Phoenix-House)

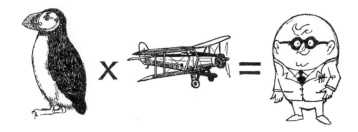

Sebastien-Roch-Nicholas Chamfort 1741–1794

38   *La Philosophie, ainsi que la Médecine, a beaucoup de drogues, très peu de bons remèdes, et presque point de spécifiques.*
Philosophy, like medicine, has plenty of drugs, few good remedies and hardly any specific cures.
Maximes et Pensées 1794, 17

Chang Chih-Tung 1837–1909

39   Chinese learning as the substance and Western learning for application.
Ch'uan-hsueh p'ien (An exhortation to learning) 1898

Pierre Teilhard de Chardin 1881–1955

40   The history of the living world can be summarised as the elaboration of ever more perfect eyes within a cosmos in which there is always something more to be seen.
The Phenomenon of Man 1959 (New York: Harper and Row) p 31

41   The earth is veiled in geometry as far back as we can see. It crystallises. But not completely.
The Phenomenon of Man 1955 (London: Collins) p 75

Erwin Chargaff 1905–

42   What counts, however, in science is to be not so much the first as the last.
Science 1971 **172** 639

43   [On the Watson–Crick style in science] That in our day such pygmies throw such giant shadows only shows how late in the day it has become.
In H F Judson *The New Yorker* 4 December 1978, p 138
See R K Merton *On the Shoulders of Giants* which is a whole book devoted to this image

44   I am not really the helpless type, but I have never been very fond of the agressive scholarship now encountered everywhere, trying to sell to humanity brand-new laws of nature as if they were used cars.
Heraclitean Fire 1978, p 128

Pierre Charron 1541–1603

45   *La vraye science et le vray étude de l'homme c'est l'homme.*
The true science and study of mankind is man.
De La Sagesse Preface

Geoffrey Chaucer ca 1340–1400

46   For out of olde feldes, as men seith,
Cometh al this newe corn fro yeer to yeer;
And out of olde bokes, in good feith,
Cometh all this newe science that men lere.
The Parlement of Foules line 22

47   Little Lewis my son, I have perceived well by certain signs thy ability to learn sciences touching numbers and proportions; and I also consider they earnest prayer specially to learn the Treatise of the Astrolabe. ... I will show thee this treatise, divided into five parts, under full easy rules and in plain English words; for Latin thou knowest as yet but little, my little son....
Treatise on the Astrolabe Preface

Anton Chekhov 1860–1904

48   Under the flag of science, art and persecuted freedom of thought Russia would oneday be ruled by toads and crocodiles the like of which were unknown even in Spain at the time of the Inquisition.
Letter of 27 August 1888 in Ronald Hingley *Russian Writers and Society* 1967 (London: Hutchinson) p 238

49   There is no national science just as there is no national multiplication table; what is national is no longer science.
In V P Ponomarev *Mysli o nauke* Kishinev, 1973, p 121

Lord Chesterfield [Philip Dormer Stanhope] 1694–1773

50   The pleasure is momentary, the position ridiculous and the expense damnable.
[Of sexual intercourse]
Nature 1970 **227** 772

51   Wear your learning, like your watch, in a private pocket; and do not pull it out and strike it, merely to show that you have one.
Letters to his son

Gilbert Keith Chesterton 1864–1936

52   It isn't that they can't see the solution. It is that they can't see the problem.
The Point of a Pin in *The Scandal of Father Brown* 1935 (London: Cassell)

53   A man must love a thing very much if he not only practices it without any hope of fame and money, but even practices it without any hope of doing it well.

54   The object of opening the mind, as of opening the mouth, is to shut it again on something solid.
In *New Scientist* 1990 **126** 72

Chou En-Lai 1898–1976

55   We must catch up with this advanced level of world science .... Only by mastering the most advanced sciences can we insure ourselves of an impregnable national defence, a powerful and up-to-date economy and adequate means to join the Soviet Union and the People's Democracies in defeating the imperial powers, either in peaceful competition or in any agressive war which the enemy may unleash.
Report on the Question of the Intellectuals 14 January 1956. *China Quarterly* 1960 35–50

56   Many of our highly skilled intellectuals have highly complicated pasts.
Report on the Question of the Intellectuals 14 January 1956

Chuang-Tze ca 369–286 BC

57   There is no bandit so remorseless as Nature. In the whole universe there is no escape from it.
Philosophy Chapter 11, transl Giles

Winston Spencer Churchill 1874–1965

58   ... I see the absolute truth and explanation of things, but something is left out which upsets the whole, so by a larger sweep of the mind I have to see a greater truth and a more complete explanation which comprises the erring element. Nevertheless, there is still something left out. So we have to take in a still wider sweep. The process continues inexorably. Depth beyond depth of unendurable truth opens.
[Describing his impressions on coming out of an anaesthetic after an accident with a taxi]
My Early Life 1930 (London: Hamlyn)

59   Although personally I am quite content with existing explosives, I feel we must not stand in the path of improvement ...
[30 August 1941. Minute on report of MAUD Committee that it would be possible to make a uranium bomb]
Statements Realting to the Atomic Bomb 1945 (London: HMSO) p 3

60   Every prophet has to come from civilisation, but every prophet has to go into the wilderness. He must have a strong impression of a complex society and all that it has to give, and then he must serve periods of isolation and meditation. This is the process by which psychic dynamite is made.
[The justification for sabbatical leave]
Essay on Moses

61   I pass with relief from the tossing sea of Cause and Theory to the firm ground of Result and Fact.
The Story of the Malakand Field Force 1898 (London: Hamlyn)

62   It is a good thing for an uneducated man to read books of quotations.
Roving Commission in *My Early Life* 1930 (London: Hamlyn) Chapter 9

63   It is by devising new weapons, and above all by scientific leadership, that
      we shall best cope with the enemy's superior strength.
      Memo for the War Cabinet 3 September 1940

64   Praise up the humanities, my boy. That will make them think that you are
      broad-minded.
      [Advice to R V Jones, his scientific consultant]
      Bulletin of the Institute of Physics 1962 **13** no 4, 101

65   Scientists should be on tap but not on top.
      [Also attributed to Walter Elliott]
      In Randolph Churchill *Twenty-one Years* (London: Weidenfeld & Nicolson) Epilogue

66   The Dark Ages may return on the gleaming wings of Science.
      [Speech at Fulton, Missouri, 5 March 1946]
      In *Science* 1969 **163** 1175

67   Unless British science had proved superior to German and unless its
      strange, sinister resources had been effectively brought to bear in the strug-
      gle for survival, we might well have been defeated, and being defeated,
      destroyed.
      The Second World War 1950 (London: Cassell) vol II, p 337

68   By God's mercy British and American science outpaced all German efforts
      .... This revelation of the secrets of nature, long mercifully witheld from
      man, should arouse the most solemn reflections in the mind and conscience
      of every human being capable of comprehension. We must indeed pray that
      these awful agencies will be made to conduce to peace among the nations,
      and that instead of wreaking measureless havoc upon the entire globe, they
      may become a perennial fountain of world prosperity.
      Statements Relating to the Atomic Bomb 1945 (London: HMSO) p 5

69   [On the Battle of Omdurman, 1898] The terrible machinery of scientific
      war had done its work. The Dervish host was scattered and destroyed. The
      end, however, only anticipates that of the victors, for Time, which laughs
      at Science, as Science laughs at Valour, will contemptuously brush both
      combatants away.
      The Morning Post 6 October 1898

70   I knew nothing of science, but I knew something of scientists, and had had
      much practice as a Minister in handling things I did not understand.
      Their Finest Hour in *War Memoirs* vol II. In *Sir Winston Churchill—a self-portrait*
      ed Colin R Coote, 1954 (London: Eyre & Spottiswoode) p 50

71   Nuclear energy is incomparably greater than the molecular energy which
      we use today .... What is lacking is the match to set the bonfire alight ....
      The Scientists are looking for this.
      Thoughts and Adventures 1932

72   ... war had been 'completely spoilt'. It is all the fault of Democracy and
      Science.
      1930
      In A Hodges *Alan Turing: the Enigma* 1983 Burnett

John Ciardi 1916–

73   Who could believe an ant in theory?
      A giraffe in blueprint?

Ten thousand doctors of what's possible
Could reason half the jungle out of being.

Marcus Tullius Cicero 106–43 BC

74   *Fortuito quodam concursu atomorum.*
By some fortuitous concourse of atoms.
[But more correctly: *Nulla cogente natura, sed concursu quodam fortuito atomorum*]
De Natura Deorum book 1, 24

75   *Omnia quae secundam Naturam fiunt sunt habenda in bonis.*
The works of Nature must all be accounted good.
De Senectute XIX, 71

76   *Nihil tam absurde dici potest, quod non dicatur ab aliquo philosophorum.*
Nothing so absurd can be said that some philosopher had not said it.
De Divinatione ii, 58

[Lord] Kenneth Clark

77   The spiritual and intellectual decline which has overtaken us in the last
thirty years ... [may be due] to the diversion of all the best brains to
technology.
Lecture on the growth of TV. In *London Review of Books* 1984 **6** no 1, p 5

Arthur Charles Clarke 1917–

78   When a distinguished but elderly scientist states that something is possible,
he is almost certainly right. When he states that something is impossible,
he is very probably wrong. (Clarke's First Law.)
Time 15 February 1971

79   How inappropriate to call this planet Earth when clearly it is Ocean.
Nature 1990 **344** 102

Paul Claudel 1868–1955

80   *Si l'ordre satisfait la raison, le désordre fait les délices de l'imagination.*
If order appeals to the intellect, then disorder titillates the imagination.
In *Structure of Non-crystalline Materials* ed P H Gaskell, 1977 (London: Taylor and
Francis) p 260

Rudolf Clausius 1822–1888

81   *Die Energie der Welt ist konstant. Die Entropie der Welt strebt einem
Maximum zu.*
The energy of the world is constant. Its entropy tends to a maximum.
1865

Georges Clemenceau 1841–1929

82   *Ce qui donne du courage, ce sont les idées.*
The thing that gives people courage is an idea.

William Kingdon Clifford 1845–1879

83   Remember, then, that scientific thought is the guide of action; that the
truth at which it arrives is not that which we can ideally contemplate
without error, but that which we may act upon without fear; and you
cannot fail to see that scientific thought is not an accompaniment of human
progress, but human progress itself.
The Common Sense of the Exact Sciences 1885 (Completed by Karl Pearson)

84 The republic of science, which allows no masters, but proved comrades only.
Mathematical Papers ed R Tucker, 1882. Reprinted 1968 (New York: Chelsea) p 561

85 An atom must be at least as complex as a grand piano.
In *D.S.B. Suppl.* 1978, p 464

86 I hold ... that in the physical world nothing else takes place but this variation [of the curvature of space].
Mathematical Papers ed R Tucker, 1882. Reprinted 1968 (New York: Chelsea) p 22

Arthur Hugh Clough 1819–1861

87 And as of old from Sinai's top
God said that God is one,
By Science strict so speaks he now
To tell us there is None.
Earth goes by chemic forces; Heaven's
A Mécanique Céleste.
And heart and mind of human kind
A watch-work as the rest.
The New Sinai

[Sir] Barnett Cocks

88 A committee is a cul-de-sac down which ideas are lured and then quietly strangled.
New Scientist 1973 **60** 424

Jean Baptiste Coffinhal fl 1794

89 *La République n'a pas besoin de Savants.*
The Republic has no need of scientists.
[Before ordering the execution of Antoine Lavoisier, May 1794]
Encyclopaedia Britannica 1911, 11th edn, 16-295d

Joel Cohen

90 Physics—envy is the curse of biology.
Science 1971 **172** 675

Simon Cohen 1894–

91   The contributions of Jews to science and invention have been directly in proportion to the amount of freedom they have enjoyed to participate in the contemporary life of the people among whom they lived.
[Director of Research, Brooklyn, New York]
Universal Jewish Encyclopedia 1943, vol 9, p 440

Samuel Taylor Coleridge 1772–1834

92   ... from the time of Kepler to that of Newton, and from Newton to Hartley, not only all things in external nature, but the subtlest mysteries of life and organisation, and even of the intellect and moral being, were conjured within the magic circle of mathematical formulae.
The Theory of Life

93   The first man of science was he who looked into a thing, not to learn whether it furnished him with food, or shelter, or weapons, or tools, or armaments, or playwiths but who sought to know it for the gratification of knowing.

94   From Avarice thus, from Luxury and War
Sprang heavenly Science; and from Science Freedom.
Religious Musings lines 224–5

95   The horned Moon, with one bright star
Within the nether tip.
[Scientists have continually reproached poets and painters for their cavalier attitude to the facts of nature]
The Ancient Mariner quoted by Coleridge from the *Phil. Trans.*

96   I should not think of devoting less than twenty years to an epic poem. Ten years to collect materials and warm my mind with universal science. I would be a tolerable mathematician. I would thoroughly understand Mechanics; Hydrostatics; Optics and Astronomy; Botany; Metallurgy; Fossilism; Chemistry, Geology; Anatomy; Medicine; then the minds of men, in all Travels, Voyages and Histories. So I would spend ten years; the next five in the composition of the poem, and the next five in the correction of it. So would I write, haply not unhearing of that divine and nightly-whispering voice, which speaks to mighty minds, of predestined garlands, starry and unwithering.
Letter to Cottle, 1796

97   Men, I think, have to be weighed, not counted.
[On the economics of the Scottish Clearances]

98   Readers may be divided into four classes: 1. Sponges, who absorb all they read, and return it nearly in the same state, only a little dirtied. 2. Sandglasses, who retain nothing, and are content to get through a book for the sake of getting through the time. 3. Strain-bags, who retain merely the dregs of what they read. 4. Mogul diamonds, equally rare and valuable, who profit by what they read, and enable others to profit by it also.
Lectures 1811–12

99   [Sir Thomas Browne discovers] quincunxes in heaven above, quincunxes in
      earth below, quincunxes in tones, in optic nerves, in roots of trees, in leaves,
      in everything.
      [What Browne actually described was hexagonal close-packing in a plane]
      Encyclopaedia Britannica 1911, 11th edn **4** 667

100   Man is so in love with intelligence, that when he is not intelligent enough
      to discover it, he will impress it.
      [On the so-called 'improvements' at Highgate]

101   We establish a centre, as it were, a sort of nucleus in the reservoir of the
      soul; and towards this, needle shoots after needle, cluster points on cluster
      points, from all parts of the contained fluid and in all directions.
      Letter, 8 April 1820

Alex Comfort 1920–

102   One has to be extraordinarily lucky, in our society, to meet one nympho-
      maniac in a lifetime.
      Darwin and the Naked Lady: Discursive Essays on Biology and Art 1962 (New York:
      Braziller) p 87

Arthur Holly Compton 1892–1962

103   The Italian Navigator has reached the New World. [And how did he find
      the Natives?] Very friendly.
      [Reporting in code by telephone to Conant that the first chain reaction had been
      initiated]
      In Laura Fermi *Atoms in the Family* 1954 (Chicago: University of Chicago Press)

Karl Taylor Compton 1887–1954

104   When I was directing the research work of students in my days at Princeton
      University, I always used to tell them that if the result of a thesis problem
      could be forseen at its beginning it was not worth working at.
      Hearings on Science Legislation USA, 1945, p 623 in D S Greenberg *The Politics of
      Pure Science* 1967 (New York: New American Library) p 115. © D S Greenberg, 1967

Auguste Comte 1798–1857

105   In mathematics we find the primitive source of rationality; and to mathematics must the biologists resort for means to carry out their researches.
Positive Philosophy transl Martineau, 5, Chapter 1

106   Men are not allowed to think freely about chemistry and biology, why should they be allowed to think freely about political philosophy?

107   *Savoir pour prévoir.*
Foreknowledge is power.

108   To understand a science it is necessary to know its history.
Positive Philosophy

James Bryant Conant 1893–1978

109   There is only one proved method of assisting the advancement of pure science—that of picking men of genius, backing them heavily, and leaving them to direct themselves.
Letter in *New York Times* 13 August 1945

110   Science owes more to the steam engine than the steam engine owes to science.
Science and Common Sense 1961 New Haven

111   Science advances, not by the accumulation of new facts, ... but by the continuous development of new concepts.

[Marquis de] Condorcet 1743–1794

112   All errors in government and in society are based on philosophic errors which, in turn, are derived from errors in natural science.
Report and Project of a Decree on the General Organisation of Public Instruction

113   *Éclairer les Sciences morales et politiques par le flambeau de l'Algèbre.*
To introduce into the moral sciences the philosophy and method of the natural sciences.

114   There should exist for all societies a science of maintaining and extending their happiness; this is what has been called *l'art social.* This science, to which all others are contributors, has not been treated as a whole. The science of agriculture, the science of economics, the science of government ... are only portions of this greater science. These separate sciences will not reach their complete development until they have been made into a well-organised whole .... And this result will be obtained sooner if all the workers are led to follow a constant and uniform method of work.
In *The validation of Scientific Theories* ed P G Frank, 1961 (New York: Macmillan) p 164

Confucius 551–479 BC

115   Tsze-kung said, 'Kwan Chung, I apprehend, was wanting in virtue. When the duke Hwan caused his brother Chiu to be killed, Kwan Chung was not able to die with him. Moreover, he became prime minister to Hwan.' The Master said, 'Kwan Chung acted as prime minister to the duke Hwan, made him leader of all the princes, and united the people enjoy the gifts

which he conferred, but for Kwan Chung, we should now be wearing our hair unbound, and the lappets of our coats buttoning on the left side.'
The Analects book 14, Chapter 18, 1–2, transl J Legge (Oxford: Oxford University Press)

116 The Master said: 'I would not have him to act with me, who will unarmed attack a tiger, or cross a river without a boat, dying without any regret. My associate must be the man who proceeds to action full of solicitude, who is fond of adjusting his plans, and then carries them into execution.'
The Analects book 7, Chapter 10, 3, transl J Legge (Oxford: Oxford University Press)

117 Study the past if you would divine the future.
The Analects

118 To no one but the Son of Heaven does it belong to order ceremonies, to fix the measures, and to determine the written characters. Now, over the kingdom, carriages have all wheels of the same size; all writing is with the same characters; and for all conduct there are the same rules.
Doctrine of the Mean Chapter 28, 2–3, transl J Legge (Oxford: Oxford University Press)

119 Learning without thinking is useless. Thinking without learning is dangerous.
Analects II, Chapter 15

120 [The Master, when he entered the temple, asked about everything. Someone questioned this behaviour. The Master said] The asking of questions is in itself the correct rite.
Analects III, Chapter 15

William Congreve 1670–1729

121 Thou art a Retailer of Phrases
And dost deal in Remnants of Remnants
Like a Maker of Pincushions
The Way of the World

John Constable 1776–1837

122 Painting is a science and should be pursued as an inquiry into the laws of nature.

William Cooper [Harry S Hoff] 1910–

123 Dibdin said: 'I see you've put your own name at the top of your paper, Mr Woods .... I always make it a matter of principle to put my name as well on every paper that comes out of the department.'
'Yours? Albert said incredulously.
'Yes,' said Dibdin, still sad and thoughtful '... and I like my name to come first—it makes it easier for purposes of identification.'
The Struggles of Albert Woods 1952 (Harmondsworth: Penguin)

Stuart A Copans

124 Why, dear colleagues, must our findings
Now be put in sterile bindings?
Once physicians wrote for recreation.
Our great teachers through the ages,

Fracastoro, and [the] other sages,
Found writing could be fun, like fornication ....
Perspectives in Biology and Medicine Winter 1973, p 232

Nicolaus Copernicus 1473–1543

125   *Mathemata mathematicis scribuntur.*
Mathematics is written for mathematicians.
De Revolutionibus Preface dedicating the book to Pope Paul III

Le Corbusier 1887–1965

126   *Une maison est une machine-à-habiter.*
A house is a machine for living in.
Vers une architecture 1923

Francis Macdonald Cornford 1874–1943

127   Greek philosophy began when Thales of Miletus successfully predicted an
eclipse of the sun in 585 BC and ended in AD 529 when the Christian Emperor Justinian closed the Schools of Athens.
Background to Modern Science 1940, ed J Needham and W Pagel (Cambridge: Cambridge University Press)

[Sir] John Cornforth 1917–

128   In a world where it is so easy to neglect, deny, pervert and suppress the
truth, the scientist may find his discipline severe. For him, truth is so seldom
the sudden light that shows new order and beauty; more often, truth is the
uncharted rock that sinks his ship in the dark.
Nobel Prize (for chemistry) Address, 1975

Paul Couderc 1899–

129   *C'est une bien faible lumière qui nous vient du Ciel étoilé. Que serait pourtant la pensée humaine si nous ne pouvions pa percevoir ces étoiles, comme
il fût arrivé par exemple si, pareille à sa soeur Vénus, la Terre s'enveloppait
toujours d'un manteau de nuages?*
The light which comes to us from the starry Heavens is very weak. But
what would human thought have been like if we could not perceive the
stars, as would have been the case if the Earth had been, like our sister
Venus, enveloped in a cloak of cloud?
Preface to *L'Architecture de l'Univers* (Paris: Gauthier-Villars)

Abraham Cowley 1618–1667

130   They chose his Eye to entertain
(His curious but not covetous Eye)
With painted Scenes and Pageants of the Brain.
Ode to the Royal Society

William Cowper 1731–1800

131   ... Some drill and bore
The solid earth, and from the strata there
Extract a register, by which we learn
That he who made it, and reveal'd its date
To Moses, was mistaken in its age.
Task Book iii *The Garden* Aldine edn, ed J Bruce, 1895 (London: Bell)

Evan Jay Crane 1889–

132   *Brevity commends itself*
      In speeches, yes
      In skirts, I guess,
      In hair, perhaps,
      In Office naps,
      And now, anew
      In abstracts too.
      [Editor of *Chemical Abstracts* 1915–1958]

Stephen Crane 1871–1900

133   A man said to the universe:
      'Sir, I exist:'
      'However,' replied the universe,
      'The fact has not created in me
      A sense of obligation.'
      War is kind and other lines 1899 (New York: Knopf)

Francis Albert Eley Crew 1886–1973

134   A few of the results of my activities as a scientist have become embedded
      in the very texture of the science I tried to serve—this is the immortality
      that every scientist hopes for. I have enjoyed the privilege, as a university
      teacher, of being in a position to influence the thought of many hundreds
      of young people and in them and in their lives I shall continue to live
      vicariously for a while. All the things I care for will continue for they will
      be served by those who come after me. I find great pleasure in the thought
      that those who stand on my shoulders will see much farther than I did in
      my time. What more could any man want?
      The Meaning of Death in *The Humanist Outlook* ed A J Ayer, 1968 (London: Pem-
      berton)

Francis Harry Compton Crick 1916–

135   If you want to understand function, study structure.
      What Mad Pursuit 1988 (London: Weidenfeld & Nicolson) p 150

136   Exact and careful model building could embody constraints that the final
      answer had in any case to satisfy.
      What Mad Pursuit 1988 (London: Weidenfeld & Nicolson) p 60

137   The job of theorists, especially in biology, is to suggest new experiments.
      A good theory makes not only predictions, but surprising predictions that
      then turn out to be true .... Let a theorist produce just one theory of the
      type sketched above and the world will jump to the conclusion (not always
      true) that he has special insight into difficult problems.
      What Mad Pursuit 1988 (London: Weidenfeld & Nicolson) p 142

138   Chance is the only source of true novelty.
      Life Itself 1982 (London: Macdonald) p 58

139   While Occam's razor is a useful tool in the physical sciences, it can be a
      very dangerous implement in biology. It is thus very rash to use simplicity
      and elegance as a guide in biological research.
      What Mad Pursuit 1988 (London: Weidenfeld & Nicolson) p 138

140   [Of René Thom] In my experience most mathematicians are intellectually
      lazy and especially dislike reading experimental papers. He seemed to have
      very strong biological intuitions but unfortunately of negative sign.
      What Mad Pursuit 1988 (London: Weidenfeld & Nicolson) p 136

141   Molecular psychology.
      What Mad Pursuit 1988 (London: Weidenfeld & Nicolson) p 163

142   It is not easy to convey, unless one has experienced it, the dramatic feeling
      of sudden enlightenment that floods the mind when the right idea finally
      clinches into place.
      What Mad Pursuit 1988 (London: Weidenfeld & Nicolson)

Francis Harry Compton Crick and John C Kendrew 1916– and 1917–

143   X-ray diffraction is not a difficult branch of physics: on the contrary, it is
      easy to the point of tediousness. The widespread view that it is unintel-
      ligible has arisen because a certain intellectual effort is needed to grasp
      its mathematical foundations, and because it is supposed, incorrectly, that
      some special type of 'three-dimensional imagination' is a prerequisite for
      understanding its methods and results.
      J Glusker and J Trueblood

Richard Howard Stafford Crossman 1907–1974

144   Isotopes ... whatever they may be.
      The Crossman Diaries quoted in *New Scientist* 28 August 1982 p 573

James Gerald Crowther 1899–

145   The State Planning Commission is the most original product of the Bol-
      shevik Revolution.
      Soviet Science London, 1936, p 15

Edward Estlin Cummings [e e cummings] 1894–1962

146   O sweet spontaneous
      earth how often
      has the naughty thumb
      of science prodded
      thy
      beauty
      thou answereth
      them only with
      spring.
      Tulips and Chimneys 1924 (New York: Seltzer)

Marie [Skłodowska] Curie 1867–1934

147   There are sadistic scientists who hasten to hunt down error instead of es-
      tablishing the truth.

148   It would be impossible, it would be against the scientific spirit .... Physi-
      cists should always publish their researches completely. If our discovery has
      a commercial future that is a circumstance from which we should not profit.
      If radium is to be used in the treatment of disease, it is impossible for us
      to take advantage of that.
      [On the patenting of radium. Discussion with her husband, Pierre]
      Eve Curie *The discovery of radium* in *Marie Curie* transl V Sheean, 1938 (London:
      Heinemann)

Nicolaus Cusanus 1401–1464

149   *Coincidentia oppositorum....*
      The identity of opposites....

[Lord] George Nathaniel Curzon 1859–1925

150   The East is a university in which the scholar never takes a degree.
      28th October 1898
      Indian Speeches I, vii. In Kenneth Rose *Superior Person* 1969 (London: Weidenfeld
      & Nicolson) p 335

151   In the case of Japan I must confess to having departed widely from the
      accepted model of treatment. There will be found nothing in those pages
      of the Japan of temples, tea-houses, and bric-a-brac—that infinitesimal
      segment of the national existence which the traveller is so prone to mistake
      for the whole, and by doing which he fills the educated Japanese with such
      unspeakable indignation. I have been more interested in the efforts of a
      nation, still in pupilage, to assume the manners of the full-grown man, in
      the constitutional struggles through which Japan is passing, in her relations
      with foreign Powers, and in the future that awaits her immense ambitions.
      Problems of the Far East 1894, Preface to the first edition

152   Scientific research is ... the apex of educational achievement.
      Budget Speech 25 March 1903

153   [German synthetic indigo is the 'blue terror' and hoped that with the help
      of researches the Bihar indigo would beat] the finest product of Teutonic
      synthesis ... [but] ... price, however, is the determining point. Science has
      got to help you to bring it down.
      [Laying the foundation-stone of the Pusa agricultural college on 1 April 1905]
      Speeches vol IV, Calcutta, 1906, pp 116–17

[Sir] Frederick Sydney Dainton 1914–

1   Accountants and second-rate business school jargon are in the ascendant. Costs, which rise rapidly, and are easily ascertained and comprehensible, now weigh more heavily in the scales than the unquantifiable and unpredictable values and future material progress. Perhaps science will only regain its lost primacy as peoples and government begin to recognise that sound scientific work is the only secure basis for the construction of policies to ensure the survival of Mankind without irreversible damage to Planet Earth.
New Scientist 3 March 1990, p 62

The Sixth Dalai Lama 1682–1705

2   This girl was perhaps not born of a mother,
But blossomed in a peach tree:
Her love fades
Quicker than peach-flowers.
Although I know her soft body
I cannot sound out her heart;
Yet we have but to make a few lines on a chart
And the distance of the farthest stars
In the sky can be measured.
Tibet transl G Tucci, 2nd edn, 1973 (London: Elek) p 173

Renato Dalbecco

3   We don't do science for the general public. We do it for each other. Good day.
Complete text of interview with H F Judson, in *The Sciences* 1983 **23** no 6, 51

Jean-le-Rond D'Alembert 1717–1783

4   *Allez en avant, et la foi vous viendra.*
Push on, and faith will catch up with you.
[Advice to those who questioned the calculus]

Salvador Dali 1904–1990

5   And now the announcement of Watson and Crick about DNA. This is for me the real proof of the existence of God.
(Playboy Magazine July 1964. In J F C Crick *Of Molecules and Men* 1966 (Washington, DC: University of Washington Press)

Cyril Dean Darlington 1903–

6   It is against the background of conflict and confusion in the relations of science and society that we find ourselves confronted with a crisis in the history of mankind, and particularly in the history of human government. It is a crisis arising from the rapidly increasing power given to man by science. It is a crisis such as we are accustomed to leave to the arbitrement of sectional interests supported by shouts and cries. But it is one to which scientific inquiry can provide a solution. For the fundamental problem of government is one that can be treated by exact biological methods. It is the problem of the character and the causation of the differences that exist among men, among the races, classes, and individuals which compose

mankind. The little passions and prejudices we have been discussing so far fade into nothingness in face of the gigantic errors and illusions that can be, and are being, mobilized to defeat or pervert scientific truth in this field.
Moncure Conway Memorial Lecture 1948 (London: Watts) p 28

7   Mankind ... will not willingly admit that its destiny can be revealed by the breeding of flies or the counting of chiasmata.
Royal Society Tercentenary Lecture 1960

8   No society can however reach a high development without some kind of priesthood, which organizes a religion designed to govern its breeding behaviour. If breeding behaviour is left at the sole discretion of a governing class, that class, and with it the whole society, is liable to disintegrate.
The Evolution of Man and Society 1969 (London: Allen & Unwin) p 675

Charles Darwin 1808–1882

9   The preservation of favourable variations and the rejection of injurious variations, I call Natural Selection, or Survival of the Fittest. Variations neither useful nor injurious would not be affected by natural select and would be left a fluctuating element.
Origin of Species 1859. 1968 (Harmondsworth: Penguin) p 131

10   Great is the power of steady misrepresentation—but the history of science shows, how fortunately, this power does not long endure.
Origin of Species 1859. 1956 (London: Dent) p 454

11   I see no good reasons why the views given in this volume should shock the religious feelings of anyone.
Origin of species 1859

12   My mind seems to have become a kind of machine for grinding general laws out of large collections of facts.
Autobiography 1859

13   Why is thought being a secretion of the brain more wonderful than gravity a property of matter?
In S J Gould *Ever since Darwin* 1978 (London: Burnett/Deutsch)

14   I had, ... during many years, followed a golden rule, namely, that whenever a published fact, a new observation or thought came across me, which was opposed by my general results, to make a memorandum of it without fail and at once; for I had found by experience that such facts and thoughts were far more apt to escape from memory than favourable ones.
The Autobiography of Charles Darwin with original omissions restored, ed Nora Barlow, 1958 (London: Collins) p 123

15   [Writing to Karl Marx, 1880] It seems to me (rightly or wrongly) that direct arguments against Christianity and Theism hardly have any effect on the public; and that freedom of thought will best be promoted by that gradual enlightening of human understanding which follows the progress of science. I have therefore always avoided writing about religion and have confined myself to science.
In S J Gould *Ever Since Darwin* 1978 (London: Burnett/Deutsch) p 26

16   False facts are highly injurious to the progress of science, for they often endure long; but false views, if supported by some evidence do little harm, for everyone takes a salutory pleasure in proving their falseness.

17   In October 1838, ... I happened to read for amusement 'Malthus on Population', and being well prepared to appreciate the struggle for existence which everywhere goes on from long-continued observation of the habits of animals and plants, it at once struck me that under these circumstances favourable variations would tend to be preserved, and unfavourable ones to be destroyed. The result of this would be the formation of new species. Here then I had at last got a theory by which to work.
Autobiography 1876

Erasmus Darwin 1731–1802

18   Shall we conjecture that one and the same kind of living filaments is and has been the cause of all organic life?
Zoonomia 1794, I, 511

19   Soon shall thy arm, unconquer'd steam: afar
Drag the slow barge, or drive the rapid car;
Or on wide-waving wings expanded bear
The flying chariot through the field of air.
The Botanic Garden I, i, 289

20   Would it be too bold to imagine that in the great lengths of time, since the earth began to exist, perhaps millions of ages before the commencement of the history of mankind, would it be too bold to imagine that all the warmblooded animals have arisen from one living filament which the Great First Cause endued with animality ... and thus possessing the faculty of continuing to improve by its own inherent activity and of delivering down those improvements by generation to its posterity, world without end.
Zoonomia 1794

21   First *heat* from chemic dissolution springs,
And gives to matter its eccentric wings;
And strong *repulsion* parts the exploding mass,
Melts into lymph, or kindles into gas.
attraction next, as earth or air subsides,
The ponderous atoms from the light divides,
Approaching parts with quick embrace combines,
Swells into spheres, and lengthens into lines.
Last, as fine goads the gluten-threads excite,
Cords grapple cords, and webs with webs unite;
And quick *contraction* with ethereal flame
Lights into life the fibre-woven frame.
The Temple of Nature Canto I, lines 223–46

[Sir] Francis Darwin 1848–1925

22   But in science the credit goes to the man who convinces the world, not to the man to whom the idea first occurs.
Eugenics Review 1914 **6** 1

Sushil Chandra Dasgupta

23   Note on Love Waves in a Homogeneous Crust Laid upon a Heterogeneous Medium.
[Title of paper]
[In fact called after A E H Love, author of *A Treatise on the Mathematical Theory of Elasticity* 1892 (Cambridge: Cambridge University Press)
Indian Journal of Theoretical Physics 1953 **1** 121

Edgerton Y Davis (Jr)

24   A drug is a substance that, when injected into a rat produces a scientific paper.
In *High Times Encyclopaedia of Recreational Drugs* 1978 (New York: Stonehill)

[Sir] Humphry Davy 1778–1829

25   The progression of physical science is much more connected with your prosperity than is usually imagined. You owe to experimental philosophy some of the most important and peculiar of your advantages. It is not by foreign conquests chiefly that you are become great , but by a conquest of nature in your own country.
Lecture at the Royal Institution 1809

26   Imagination, as well as reason is necessary to perfection in the philosophical mind. A rapidity of combination, a power of perceiving analogies, and of comparing them by facts, in the creative source of discovery. Discrimination and delicacy of sensation, so important in physical research, are other words for taste; and the love of nature is the same passion, as the love of the magnificent, the sublime and the beautiful.
Parallels between Art and Science 1807, p 208

Richard Dawkins 1941–

27   We are survival machines—robot vehicles blindly programmed to preserve the selfish molecules known as genes. This is a truth which still fills me with astonishment.
The Selfish Gene Preface

Stevan Dedijer 1911–

28   The fruitful pursuit of scientific truth and its application, once discovered, is not just a matter of talented individuals well trained in foreign universities and supplied with the equipment they desire. These are very important, but the cultivation of science is a collective undertaking [written as 'understanding'] and success in it depends on an appropriate social structure. This social structure is the scientific community and its specialised institutions.
Minerva 1963 **2** no 1, 81

John Dee 1527–1608

29   A marveilous newtrality have these things mathematicall, and also a strange participation between things supernaturall, immortall, intellectuall, simple and indivisible, and things naturall, mortall, sensible, componded and divisible.
Preface to his edition of Euclid, 1570

Daniel Defoe 1660–1731

30   ... a merchant sitting at home in his Counting-house, at once converses with all Parts of the Known World. This, and Travel, makes a True-bred Merchant the most Intelligent Man in the world, and consequently the most capable, when urg'd by Necessity, to Contrive New Ways to live.
An Essay upon Projects 1697, Introduction

31   Necessity ... has so violently agitated the wits of men at this time, that it seems not at all improper ... to call it, the Projecting Age. ... The Art of War, which I take to be the highest Perfection of Human Knowledge, is a sufficient proof of what I say, especially in conducting Armies, and in offensive Engines; witness the new ways of Mines, Fougades, Entrenchments, Attacks, Lodgments, and a long *Et Cetera* of New Inventions .... But if I would search for a Cause, from whence it comes to pass that this Age swarms with such a multitude of Projectors more than usual; who besides the Innumerable Conceptions which dye in the bringing forth ... do really every day produce new Contrivances, Engines, and Projects to get Money, never before thought of; if, I say, I would examine whence this comes to pass, it must be thus: The Losses and Depredations which this War brought with it at first were exceeding many ... [Merchants], prompted by Necessity, rack their Wits for New Contrivances, New Inventions, New Trades, Stocks, Projects, and any thing to retrieve the desperate Credit of their Fortunes.
An Essay upon Projects 1697, Introduction

Eugène Delacroix 1799–1863

32   To be a poet at twenty is to be twenty: to be a poet at forty is to be a poet.

Max Delbrück 1906–

33   While the artist's communication is linked forever with its original form, that of the scientist is modified, amplified, fused with the ideas and results of others.
In *The Eighth Day of Creation* near figure 447

J M R Delgado

34   Mind appears when symbols by material means shape the neurons.
Brain and Mind Ciba Foundation Series 69, October 1979, p 403

Democritos [of Abdera] ca 460–ca 370 BC

35   Everything existing in the Universe is the fruit of chance and necessity.
[Taken by Jacques Monod as the title of his book]
Diogenes Laertius IX

36   I am the most travelled of all my contemporaries; I have extended my field of enquiry wider than anyone else; I have seen more countries and climes, and have heard more speeches of learned men. No one has surpassed me in the composition of lines according to demonstration, not even the Egyptian knotters of ropes, or geometers.

37   Nothing exists except atoms and empty space; everything else is opinion.
Diogenes Laertius IX, 44–5

René Descartes 1596–1650

38   I thought the following four (rules) would be enough, provided that I made a firm and constant resolution not to fail even once in the observance of them.
The first was never to accept anything as true if I had not evident knowledge of its being so; that is, carefully to avoid precipitancy and prejudice, and to embrace in my judgment only what presented itself to my mind so clearly and distinctly that I had no occasion to doubt it. The second, to divide each problem I examined into as many parts as was feasible, and as was requisite for its better solution. The third, to direct my thoughts in an orderly way; beginning with the simplest objects, those most apt to be known, and ascending little by little, in steps as it were, to the knowledge of the most complex; and establishing an order in thought even when the objects had no natural priority one to another. And the last, to make throughout such complete enumerations and such general surveys that I might be sure of leaving nothing out.
These long chains of perfectly simple and easy reasonings by means of which geometers are accustomed to carry out their most difficult demonstrations had led me to fancy that everything that can fall under human knowledge forms a similar sequence; and that so long as we avoid accepting as true what is not so, and always preserve the right order of deduction of one thing from another, there can be nothing too remote to be reached in the end, or too well hidden to be discovered.
Discours de la méthode pour bien conduire sa raison et chercher la vérité dans les sciences 1637

39   If we possessed a thorough knowledge of all the parts of the seed of any animal (e.g. man), we could from that alone, by reasons entirely mathematical and certain, deduce the whole conformation and figure of each of its members, and, conversely if we knew several peculiarities of this conformation, we would from those deduce the nature of its seed.
Oeuvres iv, 494

40   It is well to know something of the manners of various peoples, in order more sanely to judge our own, and that we do not think that everything

against our modes is ridiculous, and against reason, as those who have seen nothing are accustomed to think.
Discourse on Method part 1

41   *Et ainsi nous rendre maîtres et possesseurs de la nature.*
And thus makes us the masters and possessers of nature.
Discours de la méthode pour bien conduire sa raison et chercher la vérité dans les sciences 1637

42   It is contrary to reason to say that there is a vacuum or a space in which there is absolutely nothing.
[*Natura vacuum abhorret*]
Principles of Philosophy 2
In S Sambursky (ed) *Physical Thought from the PreSocratics to the Quantum Physicists* 1974 (London: Hutchinson)

43   Which will seem in no way strange to those who, knowing how many different automata or moving-machines the industry of man can make, employing but few pieces compared with the multitude of bones, muscles, nerves, arteries, veins and all the other parts which are in the body of each animal, consider this body as a machine.
Oeuvres de Descartes ed C Adam and P Tannery, Paris, 1897–1903, vol VI, pp 51–6

44   *On ne surait rien imaginer de si étrange et de si peu croyable, qu'il n'ait été dit par quelqu'un des philosophes.*
One cannot conceive of anything so strange and so implausible that it has not already been said by one philosopher or another.
[Quoting Cicero]
Discours de la Méthode 1637, II

45   *Nihil affirmo, nihilque ab ullo credi velim, nisi quod ipse evidens et invicta Ratio persuad·bit.*
I affirm nothing and wish to believe nothing but that which is self-evident and of which invincible reason will persuade me.
Cartes. Princip part 4, art. 207

46   I think, therefore I am.
[Inscribed on his monument in Tours]

47   I think, therefore IBM.
International Business Machines advertisement, 25 January 1990

David Devant [David Wighton] 1868–1941

48   [The conjurer: of Maskelyne and Devant] You scientists are the last people likely to detect fraud. You are accustomed to deal with Nature who never cheats.
In G M Caroe *William Henry Bragg* 1978 (Cambridge: Cambridge University Press) p 102

Charles Dickens 1812–1870

49   'Yes I have a pair of eyes', replied Sam, 'and that's just it. If they was a pair o' patent double million magnifyin' gas microscopes of hextra power, p'raps I might be able to see through a flight o' stairs and a deal door; but bein' only eyes, you see, my wision's limited'.
The Pickwick Papers

50 When he has learnt that bottinney means a knowledge of plants, he goes
and knows 'em. That's our system, Nickleby; what do you think of it?
Nicholas Nickleby

Emily Dickinson 1830–1886

51 Faith is a fine invention
For gentlemen who see;
But microscopes are prudent
In an emergency.
Poems, Second Series ca 1880 XXX

Patric Dickinson 1914–

52 Who were they, what lonely men,
Imposed on the fact of night
The fiction of constellations
And made commensurable
The distances between
Themselves, their loves, and their doubt
Of governments and nations?
Who made the dark stable
When the light was not? Now
We receive the blind codes
Of spaces beyond the span
Of our myths, and a long dead star
May only echo how
There are no loves nor gods
Men can invent to explain
How lonely all men are.
Jodrell Bank in *The World I See* 1960 (London: Chatto & Windus)

Stella Didacus [Diego de Estella] 1524–1578

53 *Pygmaeos gigantum humeris impositos, plusquam ipsos gigantes videre.*
Dwarfs on the shoulders of giants see further than the giants themselves.
Eximii verbi divini CONCIONATORIS ORDINNIS MINORUM Regularis Obser-
vantiae Antwerp, 1622. See Robert K Merton *On the Shoulders of Giants* 1965 (New
York: Free Press)

Denis Diderot 1713–1784

54 ... the following definition of an animal: a system of different organic
molecules that have combined with one another, under the impulsion similar to an obtuse and muffled sense of touch given to them by the creator
of matter as a whole, until each one of them has found the most suitable
position for its shape and comfort.
On the Interpretation of Nature 1753, sec. 51

55 Do you see this egg? With it you can overthrow all the schools of theology,
all the churches of the earth.
Conversations with D'Alembert

56 We have three principal means: observation of nature, reflection, and experiment. Observation gathers the facts reflection combines them, experiment
verifies the result of the combination. It is essential that the observation

of nature be assiduous, that reflection be profound, and that experimentation be exact. Rarely does one see these abilities in combination. And so,
creative geniuses are not common.
On the Interpretation of Nature 1753, XV

57   Why should electricity not modify the formation and properties of crystals?
On the Interpretation of Nature 1753, XXXIV

58   *[Of him] Je suis bon encyclopédiste,*
Je connais le mal et le bien,
Je suis Diderot a la piste;
Je connais tout et ne crois rien.
I am the true encylopedist,
I know good and evil,
I am Diderot on the track;
I know everything and believe nothing.
In P Grosclaude *Un Audacieux Message: L'Encyclopedie* Paris, 1951, p 53

59   *Et des boyaux du dernier prêtre*
Serrons le cou du dernier roi.
And with the bowels of the last priest let us strangle the last king.
Dithyrambe sur la fête des Rois

60   *La posterité pour le philosophe, c'est l'autre monde pour*
l'homme religieux.
For the philosopher, posterity is what the after-world
is for the religious man.
Lettre à Falconet, 1765

Diogenes [the Cynic] ca 400–ca 325 BC

61   For the answer was good that Diogenes made to one that asked him in
mockery, How it came to pass that philosophers were the followers of rich
men, and not rich men of philosophers He answered soberly, and yet sharply,
Because the one sort knew what they had need of, and the other did not.
In Francis Bacon *Advancement of Learning* I, III, 10

62   [When someone asked him his nationality, he replied] cosmopolitan.
Diogenes Laertius VI, 63

Paul Adrien Maurice Dirac 1902–84

63   I think that there is a moral to this story, namely that it is more important to have beauty in one's equations than to have them fit experiment. If
Schrödinger had been more confident of his work, he could have published
it some months earlier, and he could have published a more accurate equation .... It seems that if one is working from the point of view of getting
beauty in one's equations, and if one has really a sound insight, one is on
a sure line of progress. If there is not complete agreement between the results of one's work and experiment, one should not allow oneself to be too
discouraged, because the discrepancy may well be due to minor features
that are not properly taken into account and that will get cleared up with
further developments of the theory ....
Scientific American May 1963

Milovan Djilas 1911–

64 The Party welcomed intellectuals—Tito did, and so did the war combatants—but only if they did not 'philosophise', and so long as they sacrificed themselves to the ideals of the working class—in sum, so long as they yielded to the middle segment, and eventually dissolved in it. That middle segment is suspicious of intellectuals, and especially of intellectuals who think and examine, as distinct from specialised scholars. One cannot dispense with intellectuals, and especially with 'thinkers' and 'literati'. But a certain caution toward them is essential.
Tito 1981 (London: Weidenfeld & Nicolson) p 53

Theodosius Gregorievich Dobzhansky 1900–

65 It is possible that there is, after all, something unique about man and the planet he inhabits.
Perspectives in Biology and Medicine 1972, Winter, pp 157–175

66 Nothing in biology makes sense except in the light of evolution.
Title of article in *Am. Biol. Teacher* 1973 **35**, 125–9

[Sir] William Richard Shaboe Doll 1912–

67 Basic research is not the same as development. A crash programme for the latter may be successful; but for the former it is like trying to make nine women pregnant at once in the hope of getting a baby in a month's time.
New Scientist 18 November 1976, p 375

Aelius Donatus 4th Century

68 *Pereant qui ante nos nostra dixerunt.*
To the devil with those who published before us.
Quoted by St. Jerome, his pupil

John Donne 1571–1631

69 Let me arrest thy thoughts; wonder with mee,
Why plowing, building, ruling and the rest,
Or most of those arts, whence our lives are blest,
By cursed Cains race invented be,
And blest Seth vext us with Astronomie.
There's nothing simply good, nor ill alone,
Of every quality Comparison
The onely measure is, and judge, Opinion.
The Progress of the Soule 1601, lines 513–20

70 Why grass is green, or why our blood is red
Are mysteries which none have reach'd unto.
Of the Progress of the Soul. The Second Anniversary lines 228–9

71 Then, soul, to thy first pitch work up again;
Know that all lines which circles do contain,
For once they the centre touch, do touch
Twice the circumference; and be thou such;
Double on heaven, thy thoughts on earth employed;
[*Elements of Euclid* III, 20]
Of the Progress of the Soul. The Second Anniversary lines 435–9

72   And new philosophy call all in doubt;
      The element of fire is quite put out;
      The sun is lost, and th' earth, and no man's wit
      Can well direct him where to look for it.
      And freely men confess that this world's spent
      When in the planets and the firmament
      They seek so many new and see that this
      Is crumbled out again to his atomies.
      'Tis all in pieces, all coherence gone;
      All just supply and all relation.
      Prince, subject, father, son are things forgot,
      For every man alone thinks he hath got
      To be a Phoenix, and that then can be
      None of that kind which he is but he.
      Anatomie of the World. First Anniversary 1611, lines 205–18

73   Is not thy sacred hunger of science
      Yet satisfied?
      Verse Letters: To Mr. B.B.

74   Know'st thou but how the stone doth enter in
      The bladder's cave, and never break the skin?
      Anatomie of the World. The Second Anniversarie lines 269–70

John Don Passos 1896–1970

75   All his life Steinmetz was a piece of apparatus belonging to General Electric.
      [Steinmetz invented the transformer]
      Proteus in *The 42nd Parallel* 1930 (Harmondsworth: Penguin) p 273

Fyodor Mikhailovich Dostoyevskii 1821–1881

76   Without God, all things are possible.
      [Contradicting the Biblical statement]

77   Man is pre-eminently a creative animal, predestined to strive consciously for
      an objective and to engage in engineering .... But why has he a passionate
      love for destruction and chaos also?
      Notes from the Underground 1961, transl A R MacAndrew (London: New English
      Library) p 116

Mary Douglas 1921–

78   Where there is dirt there is system. Dirt is the byproduct of a systematic
      ordering and classification of matter.
      Purity and Danger 1970 (Harmondsworth: Penguin) p 48

[Sir] Arthur Conan Doyle 1859–1930

79   You will, I am sure, agree with me that if page 534 finds us only in the
      second chapter, the length of the first one must have been really intolerable.
      Sherlock Holmes in *The Valley of Fear* Chapter 1

80   Sherlock Holmes: 'I've found it! I have found a reagent which is precipitated
      by haemoglobin and by nothing else.'
      [The first meeting of Holmes and Watson]
      A Study in Scarlet 1929 (London: Murray) p 10

81 Sherlock Holmes: 'From a drop of water a logician could predict an Atlantic or a Niagara.'
A Study in Scarlet 1929 (London: Murray) p 20

82 Sherlock Holmes: 'It is of the highest importance in the art of detection to be able to recognise out of a number of facts which are incidental and which are vital .... I would call your attention to the curious incident of the dog in the night-time.' 'The dog did nothing in the night-time.' 'That was the curious incident.'
Silver Blaze in *Memoirs of Sherlock Holmes*

83 When you have eliminated the impossible, what ever remains, however improbable, must be the truth.
The Sign of Four

84 It is a capital mistake to theorise before one has data. Insensibly one begins to twist facts to suit theories, instead of theories to suit facts.
A Scandal in Bohemia

85 'Pon my word Watson, you are coming along wonderfully. We have really done very well indeed. It is true that you have missed everything of importance, but you have hit upon the method.
A Case of Identity

86 As ... the worthy professor's stock of knowledge increased—for knowledge begets knowledge as money bears interest—much of which had seemed strange and unaccountable began to take another shape in his eyes. New trains of reasoning became familiar to him, and he perceived connecting links where all had been incomprehensible and startling.
The Great Keinplatz Experiment 1894

John William Draper 1811–1882

87 How is it that the Church produced no geometers in her autocratic reign of twelve hundred years?
The Conflict Between Science and Religion 1890 (London: Kegan Paul, Trench, Trübner) p 307

John Dryden 1631–1700

88 [Of George Villiers, second Duke of Buckingham who 'made the whole body of vice his study'] A man so various that he seem'd to be
Not one, but all mankind's epitome.
Stiff in opinions, always in the wrong;
Was everything by starts, and nothing long:
But, in the course of one revolving moon,
Was chemist, fiddler, statesman, and buffoon.

Absalom and Achitophel I, 545

89   Mere poets are sottish as mere drunkards are, who live in a continual mist,
without seeing or judging anything clearly. A man should be learned in
several sciences, and should have a reasonable, philosophical and in some
measure a mathematical head, to be a complete and excellent poet.
Notes and Observations on The Empress of Morocco 1674

Albrecht Dürer 1471–1528

90   But when great and ingenious artists behold their so inept performances,
not undeservedly do they ridicule the blindness of such men; since sane
judgment abhors nothing so much as a picture perpetrated with no technical knowledge, although with plenty of care and diligence. Now the sole
reason why painters of this sort are not aware of their own error is that they
have not learnt Geometry, without which no one can either be or become
an absolute artist; but the blame for this should be laid upon their masters,
who are themselves ignorant of this art.
The Art of Measurement 1525. Preface to *Of the Just Shaping of Letters* transl R T
Nichol, book III

91   *Welcher aber ... durch die Geometria sein Ding beweist und die gründliche
Wahrheit anzeigt, dem soll alle Welt glauben. Denn da ist man gefangen.*
Whoever ... proves his point and demonstrates the prime truth geometri-
cally should be believed by all the world, for there we are captured.
Von menschlicher Proportion in *Albrecht Dürer's schriftlicher Nachlass* ed J Hei-
drich, Berlin, 1920, p 270

Pierre Duhem 1861–1916

92   This whole theory of electrostatics constitutes a group of abstract ideas
and general propositions, formulated in the clear and precise language of
geometry and algebra, and connected with one another by the rules of strict
logic. This whole fully satisfies the reason of a French physicist and his
taste for clarity, simplicity and order .... Here is a book [by Oliver Lodge]
intended to expound the modern theories of electricity and to expound a
new theory. In it are nothing but strings which move around pulleys, which
roll around drums, which go through pearl beads ... toothed wheels which
are geared to one another and engage hooks. We thought we were entering
the tranquil and neatly ordered abode of reason, but we find ourselves in
a factory.
La Théorie Physique Chapter 4, 5. In M B Hesse *Models and Analogies in Science*
1963 (London: Sheed and Ward) p 2

Freeman Dyson 1923–

93   Most of the papers which are submitted to the *Physical Review* are rejected,
not because it is impossible to understand them, but because it is possible.
Those which are impossible to understand are usually published.
Innovation in Physics

Maria von Ebner-Eschenbach 1830-1916

1 The manuscript in the drawer either rots or ripens.
Aphorismen

[Meister] Eckhart ca 1260–1327

2 The greatest power available to man is not to use it.
Nature 1973 **245** 279

Umberto Eco 1932–

3 Indeterminacy, complementarity, non-causality are not modes of being in the physical world, but systems for describing it in a convenient way.
The Role of the Reader Bloomington, IN, 1979, p 66

[Sir] Arthur Stanley Eddington 1882–1944

4 I believe there are 15,747,724,136,275,002,577,605,653,961,181,555,468, 044,717,914,527,116,709,366,231,425,076,185,631,031,296 protons in the universe and the same number of electrons.
$[2^{256} \times 136]$
Tarner Lecture 1938. In *The Philosophy of Physical Science* 1939 (Cambridge: Cambridge University Press) p 170

5 There was once a brainy baboon,
Who always breathed down a bassoon,
For he said, 'It appears
That in billions of years
I shall certainly hit on a tune.'
New Pathways in Science 1935 (Cambridge: Cambridge University Press) Chapter 3

6 We used to think that if we knew one, we knew two, because one and one are two. We are finding that we must learn a great deal more about 'and'.

7 It is also a good rule not to put overmuch confidence in the observational results that are put forward until they are confirmed by theory.
In II Judson *New Yorker* 4 December 1978, p 132

8 We have found a strange footprint on the shores of the unknown. We have devised profound theories, one after the other, to account for its origin. At last, we have succeeded in reconstructing the creature that made the footprint. And lo! It is our own.

9 Schrödinger's wave-mechanics is not a physical theory, but a dodge—and a very good dodge too.
Nature of the Physical World 1928 (Cambridge: Cambridge University Press) p 219

[Sir] Anthony Eden 1897–1977

10 Every succeeding scientific discovery makes greater nonsense of old-time conceptions of sovreignty.
Speech in the House of Commons, 22 November 1945

Gösta Carl Henrik Ehrensvard 1905–

11 Consciousness will always be one degree above comprehensibility.
Man on Another World 1965 (Chicago: University of Chicago Press)

Paul Ehrlich 1854–1915

12   Success in research needs four Gs: *Glück, Geduld, Geschick und Geld.*
Luck, patience, skill and money.
In M Perutz *Nature* 1988 **332** 791

Paul Ralph Ehrlich 1932–

13   The first rule of intelligent tinkering is to save all the parts.
Saturday Review 5 June 1971

Manfred Eigen 1927–

14   A theory has only the alternative of being right or wrong. A model has a
third possibility: it may be right, but irrelevant.
The Physicist's Conception of Nature 1973, ed Jagdish Mehra (Dordrecht: Reidel)
p 618

Albert Einstein 1879–1955

15   Commonsense is nothing more than a deposit of prejudices laid down by
the mind before you reach eighteen.
In E T Bell *Mathematics, Queen and Servant of the Sciences* 1952 (London: Bell)
p 42

16   Dear Sir   April 23, 1953

Development of Western Science is based on two great achievements; the
invention of the formal logical system (in Euclidean geometry) by the Greek
philosophers, and the discovery of the possibility to find out causal relationship by systematic experiment (Renaissance). In my opinion one has
not to be astonished that the Chinese sages have not made these steps. The
astonishing thing is that these discoveries were made at all.

Sincerely yours,

A Einstein
Letter to J E Switzer in D J de S Price *Science since Babylon* 1962 (New Haven, CT:
Yale University Press)

17   God does not care about our mathematical difficulties. He integrates empirically.
In L Infeld *Quest* 1942 (London: Gollancz) p 222

18   *Gott würfelt nicht.*
God casts the die, not the dice (Jean Untermeyer).
Albert Einstein: Creator and Rebel 1973 (London: Hart-Davis) p 193

19   The history of scientific and technical discovery teaches us that the human
race is poor in independent thinking and creative imagination. Even when
the external and scientific requirements for the birth of an idea have long
been there, it generally needs an external stimulus to make it actually
happen; man has, so to speak, to stumble right up against the thing before
the right idea comes.

20   How is it possible to control man's mental evolution so as to make him proof against the psychoses of hate and destructiveness. Here I am thinking by no means only of the so-called uncultured masses. Experience proves that it is rather the so-called Intelligentzia that is most apt to yield to these disastrous collective suggestions, since the intellectual has no direct contact with life in the raw but encounters it in its easiest synthetic form—the printed page.
Letter to Sigmund Freud

21   If you are out to describe the truth, leave elegance to the tailor.
[On being reproached that his formula of gravitation was longer and more cumbersome than Newton's]

22   Imagination is more important than knowledge.
On Science

23   Most people say that it is the intellect which makes a great scientist. They are wrong: it is the character.

24   Not only to know how nature is and how her transactions are carried through, but also to reach as far as possible the utopian and seemingly arrogant aim of knowing why nature is thus and not otherwise ....
Festschrift für Aurel Stodola 1929 (Zürich: Orell Füssli) p 126

25   One thing I have learned in a long life: that all our science, measured against reality, is primitive and childlike—and yet it is the most precious thing we have.
Albert Einstein: Creator and Rebel 1973 (London: Hart-Davis) p vii

26   The only justification for our concepts is that they serve to represent the complex of our experiences; beyond this they have no legitamacy. I am convinced that the philosophers had a harmful effect upon the progress of

scientific thinking in removing certain fundamental concepts from the domain of empiricism, where they are under control, to the intangible heights of the *a priori*—the universe of ideas is just as little independent of the nature of our experiences as cloths are of the form of the human body.
In P A Schlipp *Albert Einstein: Philosopher–Scientist* 1951, 2nd edn (New York: Tudor)

27   Our age is characterized by perfecting the means, while confusing the goals.

28   *Raffiniert ist der Herr Gott, aber boshaft ist er nicht.*
God is subtle but he is not malicious.
[Note in the Professor's lounge of the Mathematics Department at Princeton. 'God is slick, but he ain't mean'—Einstein's own translation to Derek Price, 1946]

29   The Temple of Science is a multi-faceted building.
In G Holton *Thematic Origins of Scientific Thought* 1973 (Cambridge, MA: Harvard University Press) p 376

30   When I study philosophical works I feel I am swallowing something which I don't have in my mouth.

31   The human mind has first to construct forms, independently, before we can find them in things.
In S Chandrasekhar *Nature* 1990 **344** 285

32   I cannot seriously believe in [the quantum theory] because it cannot be reconciled with the idea that physics should represent a reality in time and space, free from spooky actions at a distance [*spukhafte Fernwirkungen*].
Einstein to Born, March 1947

33   Not everyone is as fortunate as Christ. To sacrifice yourself and do some good, that takes luck.
In *Szilard*, p 12

34   [Newton wrote to Halley ... that he would not give Hooke any credit] That, alas, is vanity. You find it in so many scientists. You know, it has always hurt me to think that Galileo did not acknowledge the work of Kepler.
Einstein—A Centenary Volume ed A P French, 1979 (London: Heinemann) p 41

35   The psychical entities which seem to serve as elements in thought are certain signs and more or less clear images which can be 'voluntarily' reproduced and combined .... But taken from a psychological viewpoint, this combinatory play seems to be the essential feature in productive thought— before there is any connection with logical construction in words or other kinds of signs which can be communicated to others. The above-mentioned elements are, in my case, of visual and some muscular type. Conventional words or other signs have to be sought for laboriously only in a secondary stage, when the mentioned associative play is sufficiently established and can be reproduced at will.
Letter to Jacques Hadamard in *Einstein. A Centenary Volume* 1979 ed A P French (London: Heinemann) p 156

36   I learned many years ago never to waste time trying to convince my colleagues.
In R G Colodny (ed) *Logic, Laws and Life* 1977 (Pittsburg, PA: Pittsburg University Press) p 183

37   [Asked about a book in which 100 Nazi professors charged him with scientific error] Were I wrong, one professor would have been quite enough.
Quoted by Daniel Greenberg *Washington Post* 12 December 1978

38   [After Heisenberg's 1927 lecture enunciating the uncertainty principle] Marvellous, what ideas the young people have these days. But I don't believe a word of it.
In W Ehrenberg *Physics Bulletin* 1979 **30** 262

39   What I am really interested in is whether God could have made the world in a different way; that is, whether the necessity of logical simplicity leaves in freedom at all.
To Ernst Strauss, in G Holton *The Scientific Imagination: Case Studies* 1978 (Cambridge: Cambridge University Press)

40   To punish me for my contempt for authority, Fate made me an authority myself.
In Banesh Hoffman *Albert Einstein: Creator and Rebel* 1972 (New York: Viking) p 24

41   ... concerning the situation of scientists in America. Instead of trying to analyse the problem, I should like to express my feeling in a short remark: if I were a young man again and had to decide how to make a living, I would not try to become a scientist or scholar or teacher. I would rather choose to be a plumber or a peddler, in the hope of finding that modest degree of independence still available under present circumstances.
The Reporter 18 November 1954

42   You believe in a God who plays dice, and I in complete law and order in a world which objectively exists, and which I, in a wildly speculative way, am trying to capture .... Even the great initial success of the quantum theory does not make me believe in the fundamental dice game, although I am well aware that your younger colleagues interpret this as a consequence of senility.
Letter to Max Born

Dwight David Eisenhower 1890–1969

43   In the councils of government we must guard against the acquisition of unwarranted influence, whether sought or unsought, by the military-industrial complex. The potential for the disastrous rise of misplaced power exists and will persist .... In holding scientific research and discovery in respect, as we should, we must be alert to the equal and opposite danger that public policy could itself become the captive of a scientific–technological elite.
Farewell Address as President of the USA, 1961

44   It was not necessary to hit them with that awful thing.
[The atomic bomb on Hiroshima]

Walter Elliot 1888–1958

45   Scientific men should be on tap but not on top.
The Prof in Two Worlds 1961 (London: Collins) p 192

46   Force is not to be used to its uttermost. Nor is thought to be pushed to its logical conclusion.
Address to the General Assembly of the Church of Scotland *The Scotsman* 22 May 1957

Ralph Waldo Emerson 1803–1882

47   The English mind turns every abstraction it can receive into a portable utensil, or working institution.
Sci. Policy 1973 **2** 142

48   A foolish consistency is the hobgoblin of little minds, adored by little statesmen and philosophers and divines.
Self-reliance Essay

49   I hate quotations. Tell me what you know.
Journals May 1849

50   If a man ... make a better mousetrap than his neighbour, tho' he build his house in the woods, the world will make a beaten path to his door.
[See the *Oxford Dictionary of Quotations* for the quotation's history]

51   An institution is the lengthened shadow of one man.
Self-reliance

52   It is as difficult to appropriate the thoughts of others as to invent.

53   Things are in the saddle
And ride mankind.
Ode, inscribed to W H Channing

54   Tobacco, coffee, alcohol, hashish, prussic acid, strychnine, are weak dilutions; the surest is time. This cup which nature puts to our lips, has a wonderful virtue, surpassing that of any other draught. It opens the senses, adds power, fills us with exalted dreams which we call hope, love, ambition, science; especially it creates a craving for larger draughts of itself.
Society and Solitude in *Old Age*

55   For the world was built in order,
And the atoms march in tune.
Monadnoc

Friedrich Engels 1820–1895

56   ... science progresses in proportion to the mass of knowledge that is left to it by preceding generations, that is under the most ordinary circumstances in geometrical proportion.
Umrisse zu einer Kritik der Nationalökonomie K Marx–F Engels Werke, Band 1, Berlin, 1961, p 521

57   Freedom is the recognition of necessity.
[The title chosen by J D Bernal for his book of essays]

58   If you think that I have got hold of something here please keep it to yourself. I do not want some lousy Englishman to steal the idea. And it will take a long time to get it into shape.
Letter to Marx in *Selected Writings* ed W O Henderson, 1967 (Harmondsworth: Penguin) p 395

59   In science, each new point of view calls forth a revolution in nomenclature.
Marx–Engels Collected Works 2nd Russian edn, vol 23, p 31

60   In the book [*A Treatise on Natural Philosophy* Oxford, 1867] by these two Scotsmen [W Thomson and P G Tait] thinking is forbidden, only calculation is permitted. No wonder that at least one of them, Tait, is accounted one of the most pious Christians in pious Scotland.
Dialectics of Nature 1954 (Moscow: FLPH) p 124

61   Just as Darwin discovered the law of evolution in organic nature, so Marx discovered the law of evolution in human history; he discovered the simple fact, hitherto concealed by an overgrowth of ideology, that mankind must first of all eat and drink, have shelter and clothing, before it can pursue politics, science, religion, art, etc, and that therefore the production of the immediate material means of subsistence and consequently the degree of economic development attained by a given people or during a given epoch, form the foundation upon which the state institutions, the legal conceptions, the art and even the religious ideas of the people concerned have been evolved, and in the light of which these things must therefore be explained, instead of *vice versa* as had hitherto been the case. But that is not all. Marx also discovered the special law of motion governing the present-day capitalist mode of production and the bourgeois society that this mode of production has created. The discovery of surplus value suddenly threw light on the problem in trying to solve which all previous investigators, both bourgeois economists and socialist critics, had been groping in the dark. Two such discoveries would be enough for one lifetime. Happy the man to whom it is granted to make even one such discovery. But in every single field which Marx investigated—and he investigated very many fields, none of them superficially—in every field, even in that of mathematics, he made independent discoveries. Such was the man of science. But this was not even half the man. Science was for Marx a historically dynamic, revolutionary force. However great the joy with which he welcomed a new discovery in some theoretical science whose practical application perhaps it was as yet quite impossible to envisage, he experienced quite another kind of joy when the discovery involved immediate revolutionary changes in industry and in the general course of history.
[Funeral oration]
Marx–Engels Selected Works vol 2, p 153f

62   Life is the mode of existence of proteins, and this mode of existence essentially consists in the constant self-renewal of the chemical constituents of these substances.
Anti-Dühring 1878

63   While natural science up to the end of the last century was predominantly a *collecting* science, a science of finished things, in our century it is essentially a *classifying* science, a science of processes, of the origin and development of these things and of the interconnection which binds all these processes into one great whole.
Ludwig Feuerbach 1886

64   One day we shall certainly 'reduce' thought experimentally to molecular and chemical motions in the brain; but does that exhaust the essence of thought?
Dialectics of Nature

65   Without analysis, no synthesis.
Anti-Dühring 1878

66   Of course we still have insufficient technicians, agronomists, engineers, chemists, architects, etc, but if the worst comes to the worst we can buy them for ourselves just as the capitalists do, and if some of them turn out to be traitors (as will certainly be the case in that society) they will be punished for the edification of the others so that they will appreciate that it is not in their own interests to rob us any more. But with the exception of these specialists, and also the school teachers, we can very well do without the rest of the 'educated'.
Quoted in *Nature* 1978 **276** 146

Dennis Joseph Enright 1920–

67   To shoot a man against the National Library wall
The East unsheathes its barbarous finger-nail.
In Europe this was done in railway trucks,
Cellars underground, and such sequestered nooks.
[Written while visiting professor in Bangkok]
An Unfortunate Poem (II) Warm Protest in *Addictions* 1962 (London: Chatto & Windus) p 16

Epicurus 341–270 BC

68   If we were not troubled by doubts about the heavens, and about the possible meaning of death, and by the failure to understand the limits of pain and desire, then we should have no need of natural philosophy.
Kyriai doxai 11

69   The atoms come together in different order and position, like the letters, which, though they are few, yet, by being placed together in different ways, produce innumerable words.
Lactantius, Divin. Inst. book 3, Chapter 19. In Max Muller *Science of Language* 1871, II, p 81

Euclid [of Alexandria] ca 365 BC

70   A youth who had begun to read geometry with Euclid, when he had learnt the first proposition, inquired, 'What do I get by learning these things?' So

Euclid called a slave and said 'Give him threepence, since he must make a gain out of what he learns.'
In Stobaeus *Extracts*

Leonard Euler 1707–1783

71 Madam, I have just come from a country where people are hanged if they talk.
[In Berlin, excusing his taciturnity in conversation with the Queen Mother of Prussia, on his return from Russia]
In A Vucinich *Science in Russian Culture* 1965 (London: Peter Owen)

Euripides ca 484–ca 406 BC

72 Mighty are numbers, and joined with art resistless. [or] There is power in numbers and cunning makes us strong.
Hecuba line 884

Edward Evan Evans-Pritchard 1902–1973

73 Get yourself a decent hamper from Fortnum and Mason's and keep away from the native women.
[British social anthropologist giving his advice on field-work]
In Nigel Barley *The Innocent Anthropologist* The British Museum, 1983

Henri Jean Fabre 1823–1915

1   History celebrates the battlefields whereon we meet our death, but scorns to speak of the ploughed fields whereby we live. It knows the names of the kings' bastards, but cannot tell us the origin of wheat.

2   *S'il est un légume du bon Dieu sur la terre, c'est bien l'haricot.*
If there is one vegetable which is God-given, it is the haricot bean.
Souvenirs Entomologiques 8me ser. IV, La Bruche des Haricots, Delagrave

John King Fairbank 1907–

3   The question now is: Can we understand our stupidity? This is a test of intellect, not of character.
[Professor of Chinese at Harvard]
The Observer 4 May 1975, p 10

Fang Li-zhi 1936

4   Intellectuals, who own and create information and knowledge are the most dynamic component of the productive forces; this is what determines their social status.
[Vice-president, Hefei University]
The Beijing Review ca December 1986. In *The Guardian* 5 February 1987

Michael Faraday 1791–1867

5   [On being offered the Presidency of the Royal Society] Tyndall, I must remain plain Michael Faraday to the last; and let me now tell you, that if I accepted the honour which the Royal Society desires to confer upon me, I would not answer for the integrity of my intellect for a single year.
Tyndall's life in *Experimental Researches in Electricity* (London: Dent) p xiii

6   One day, Sir, you may tax it.
[To Mr Gladstone, the Chancellor of the Exchequer, who asked about the practical worth of electricity]
In R A Gregory *Discovery* (London: Macmillan) p 3

7   Work, Finish, Publish.
[Benjamin Franklin said much the same]

8   The world little knows how many of the thoughts and theories which have passed through the mind of a scientific investigator have been crushed in silence and secrecy by his own severe criticism and adverse examination; that in the most successful instances not a tenth of the suggestions, the hopes, the wishes, the preliminary conclusions have been realised.

9   It is the great beauty of our science that advancement in it, whether in a degree great or small, instead of exhausting the subject of research, opens the doors to further and more abundant knowledge, overflowing with beauty and utility.

Haneef A Fatmi and Robert W Young

10   Intelligence is that faculty of mind, by which order is perceived in a situation previously considered disordered.
Nature 1970 **228** 97

Gustav Theodor Fechner 1801–1887

11   *Vergleichende Anatomie der Engel.*
On the comparative anatomy of angels.
Book title, 1825 (under the pseudonym of Dr Mises)

Evgraf Stepanovich Fedorov 1853–1919

12   Crystallisation is death.
Perfectionism 1906, Izv. Peterburgskoi biologicheskoi laboratorii **8** sec. 1, 25–65, sec. 2, 9–65

Feng-shen Yin-Te 1771–1810

13   With a microscope you see the surface of things. It magnifies them but does not show you reality. It makes things seem higher and wider, But do not suppose you are seeing things in themselves.
The Microscope 1798. See *Report of the Librarian of Congress* 1937, p 177

Ernest Francisco Fenollosa 1853–1908

14   The forces which produce the branch angles of an oak lay potent in the acorn.
1904, in Hugh Kenner *The Pound Era* 1972 (London: Faber) p 163

Lawrence Ferlinghetti 1919–

15   Constantly risking absurdity
and death
whenever he performs
above the heads
of his audience
the poet like an acrobat
climbs on rime
to a high wire of his own making
... For he is a super realist
who must perforce perceive
taut truth
before the taking of each stance or step
in his supposed advance
towards that still higher perch
where beauty stands and waits
with gravity
to start her death-defying leap.
A Coney Island of the Mind 1958 (New York: New Directions) © Lawrence Ferlinghetti, 1958

Pierre de Fermat 1601–1665

16   [In the margin of his copy of Diophantus' *Arithmetica* Fermat wrote] To divide a cube into two other cubes, a fourth power or in general any power whatever into two powers of the same denomination above the second is impossible, and I have assuredly found an admirable proof of this, but the margin is to narrow to contain it.
[This was his famous *Last Theorem*]
Translated from Latin in *Source Book of Mathematics* 1929 (New York: McGraw-Hill)

Enrico Fermi 1901–1954

17   Whatever Nature has in store for mankind, unpleasant as it may be, men must accept, for ignorance is never better than knowledge.
In Laura Fermi *Atoms in the Family* 1954 (Chicago: University of Chicago Press)

18   A general is a man who takes chances. Mostly he takes a fifty–fifty chance; if he happens to win three times in succession he is considered a great general.
In *Szilard*, p 147

19   Before I came here I was confused about this subject. Having listened to your lecture I am still confused. But on a higher level.

Jean Fernel 1497–1558

20   [Life is] a low, flame-less fire.
In C S Sherrington *The Endeavour of Jean Fernel* 1946 (Toronto: MacMillan)

Ludwig Feuerbach 1804–1872

21   To bathe is the first, though the lowest, of virtues. In the stream of water the fever of selfishness is allayed. Water is the readiest means of making friends with Nature. Man rising from water is new, regenerate man.
In J Needham *Moulds of Understanding* 1976 (London: Allen & Unwin) p 32

Paul Feyerabend 1924–

22   Unanimity of opinion may be fitting for a church, for the frightened or greedy victims of some (ancient or modern) myth, or for the weak and willing followers of some tyrant. Variety of opinion is necessary for objective knowledge.
Against Method p 46

Richard Phillips Feynman 1918–88

23   Everything is made of atoms. That is the key hypothesis. The most important hypothesis in all of biology, for example, is that everything that animals do, atoms do. In other words, there is nothing that living things do that cannot be understood from the point of view that they are made of atoms acting according to the laws of physics. This was not known from the beginning it took some experimenting and theorizing to suggest this hypothesis, but now it is accepted, and it is the most useful theory for producing new ideas in the field of biology.
The Feynman Lectures on Physics vol 1, 1963 (San Francisco: Addison-Wesley) pp 1–8

24   The whole question of imagination in science is often misunderstood by people in other disciplines. They try to test our imagination in the following way. They say, 'Here is a picture of some people in a situation. What do you imagine will happen next' When we say, 'I can't imagine,' they may think we have a weak imagination. They overlook the fact that whatever we are allowed to imagine in science must be *consistent with everything else we know*; that the electric fields and the waves we talk about are not just some happy thoughts which we are free to make as we wish, but ideas which must be consistent with all the laws of physics we know. We can't allow ourselves to seriously imagine things which are obviously in contradiction

to the known laws of nature. And so our kind of imagination is quite a difficult game. One has to have the imagination to think of something that has never been seen before, never been heard of before. At the same time the thoughts are restricted in a straitjacket, so to speak, limited by the conditions that come from our knowledge of the way nature really is. The problem of creating something which is new, but which is consistent with everything which has been seen before, is one of extreme difficulty.

The Feynman Lectures on Physics vol 2, 1963 (San Francisco: Addison-Wesley) 20–11

25   We have a habit in writing articles published in scientific journals to make the work as finished as possible, to cover up all the tracks, to not worry about the blind alleys or describe how you had the wrong idea first, and so on. So there isn't any place to publish, in a dignified manner, what you actually did in order to get to do the work.

Nobel Lecture 1966

26   When I was at Princeton in the 1940s I could see what happened to those great minds at the Institute for Advanced Study, who had been specially selected for their tremendous brains and were now given this opportunity to sit in this lovely house by the woods there, with no classes to teach, with no obligations whatsoever. These poor bastards could now sit and think clearly all by themselves, OK? So they don't get an idea for while. They have every opportunity to do something, and they're not getting any ideas. I believe that in a situation like this a kind of guilt or depression worms inside of you, and you begin to *worry* about not getting any ideas. And nothing happens, still no ideas come.

Nothing happens because there's not enough *real* activity and challenge. You're not in contact with the experimental guys. You don't have to think how to answer questions from students. Nothing.

27   It seems to me that what can happen in the future is ... that the experiments get harder and harder to make, more and more expensive ... and it [scientific discovery] gets slower and slower.

The Character of Physical Law Cambridge, MA, 1965, p 172

28   Reality must take precedence over public relations, for nature cannot be fooled.

Final Report of the Inquiry into the Challenger disaster of January 1986

29   We have always had a great deal of difficulty understanding the world view that quantum mechanics represents.

At least I do, because I'm an old enough man that I haven't got to the point that this stuff is obvious to me.

Okay, I still get nervous with it....

You know how it always is, every new idea, it takes a generation or two until it becomes obvious that there's no real problem.

I cannot define the real problem, therefore I suspect there's no real problem, but I'm not sure there's no real problem.

Int. J. Theor. Phys. 1982 **21**, 471, in N D Mermin *Physics Today* April 1985, 2–21

[Sir] Ronald Aylmer Fisher 1890–1962

30   Natural selection is a mechanism for generating an exceedingly high degree of improbability.

31   No aphorism is more frequently repeated ... than that we must ask Nature few questions, or ideally, one question at a time. The writer is convinced that this view is wholly mistaken. Nature, he suggests, will best respond to a logically and carefully thought out questionnaire; indeed if we ask her a single question, she will often refuse to answer until some other topic has been discussed.
      Perspectives in Medicine and Biology 1973, Winter p 180

32   To call in the statistician after the experiment is done may be no more than asking him to perform a postmortem examination: he may be able to say what the experiment died of.
      Indian Statistical Congress, Sankhya, ca 1938

Michael Flanders 1922–1975

33   One of the great problems of the world today is undoubtedly this problem of not being able to talk to scientists, because we don't understand science; they can't talk to us because they don't understand anything else, poor dears.
      At the Drop of a Hat ca 1964

Gustave Flaubert 1821–1880

34   Poetry is as exact a science as geometry.

James Elroy Flecker 1884–1915

35   Hassan: I shall study the reasons for the excessive ugliness of the pattern of this carpet.
      Hassan

36   Caliph: Ah, if there shall ever arise a nation whose people have forgotten poetry or whose poets have forgotten the people, though they send their ships round Taprobane and their armies across the hills of Hindustan, though their city be greater than Babylon of old, though they mine a league into the earth or mount to the stars on wings—what of them?
      Hassan: They will be a dark patch upon the world.
      Hassan

John Ambrose Fleming 1849–1945

37   [Professor of electrical engineering at University College, London]
      This war is a war quite as much of chemists and engineers as of soldiers and sailors.
      Nature 1915–16 **96** 180–5
      [But, in the ten months which had passed since the war had begun, he had never been asked] to serve on any committee, cooperate in any experimental work, or place expert knowledge, which it has been the work of a lifetime to obtain, at the disposal of the forces of the Crown.
      Letter to *The Times* 15 June 1915

Bernard Le Bovier [Sieur de] Fontenelle 1657–1717

38   *Toute la philosophie n'est fondée que sur deux choses: sur ce qu'on a l'esprit curieux et les yeux mauvais.*
      Science originates from curiosity and weak eyes.
      Entretiens sur la Pluralité des Mondes Habités, Premier Soir

39 A work of morality, politics, criticism ... will be more elegant, other things being equal, if it is shaped by the hand of geometry.
Preface sur l'Utilité des Mathématiques et de la Physique 1729

40 Mathematicians are like lovers .... Grant a mathematician the least principle, and he will draw from it a consequence which you must also grant him, and from this consequence another.

41 When the heavens were a little blue arch, stuck with stars, methought the universe was too straight and close: I was almost stifled for want of air: but now it is enlarged in height and breadth, and a thousand vortices taken in. I begin to breathe with more freedom, and I think the universe to be incomparably more magnificent than it was before.
[Written in 1686]

42 There is an order which regulates our progress. Every science develops after a certain number of preceding sciences have been developed, and only then; it has to await its turn to burst its shell.
Préface des élémens de la géométrie de l'infinie in *Oeuvres* Paris, 1790, vol X, p 40

Edward Morgan Forster 1879–1970

43 The implacable offensive of science.

44 [Attacking J D Bernal] Owing to the political needs of the moment, the scientist occupies an abnormal position, which he tends to forget. He is subsidised by the terrified governments who need his aid, pampered and sheltered as long as he is obedient, and prosecuted under the Official Secrets Act when he has been naughty. All this separates him from ordinary men and women and makes him unfit to enter into their feelings. It is high time he came out of his ivory laboratory. We want him to plan for our bodies. We do not want him to plan for our minds.
The Challenge of Our Time radio broadcast, 1946

Michel Paul Foucault 1926–1988

45 All modern thought is permeated by the idea of thinking the unthinkable.

Jean Fourastié 1907–1990

46 *Le rétard des sciences économiques et sociales sur les sciences de la matière est l'une des causes des malheurs actuels de l'homme vers des horizons imprévus.*
The backwardness of the economic and social sciences with respect to the sciences of matter is one of the causes of the present human calamities through lack of foresight.
Le Grand Espoir de XXme Siècle (Paris: Gallimard) Introduction

Antoine François de Fourcroy 1755–1809

47 *Mort au Tyrans. Programmes des cours révolutionnaires sur la fabrication des salpêtres, des poudres et des canons. Faits a Paris, par ordre du Comité de Salut public, dans l'amphithéatre du Museum d'Histoire Naturelle, et dans la salle des Electeurs, maison du ci-devant Évêché, les 1, 11 et 12 ventose, deuxième année de la République française une et indivisible, par les citoyens Guyton, Fourcroy, Dufourny, Berthollet, Carny, Pluvinet, Monge, Hassenfratz et Perrier.*

Death to tyrants. Programmes of revolutionary courses on the making of gunpowder and of cannons. Delivered in Paris, by order of the Committee for Public Safety, in the amphitheatre of the Natural History Museum, and in the Hall of Electors, the house of the former Bishop on the first, eleventh and twelfth of 'ventose', in the second year of the United French Republic, by citizens Guyton, Fourcroy, Dufourny, Berthollet, Carny, Pluvinet, Monge, Hassenfratz and Perrier.
14 lectures each of 2 or 4 pages, 1794

George Fox 1624–1691

48   I came to know God experimentally.
[Fox was the founder of the Society of Friends]

[Sir] Theodore Fox 1899–

49   We shall have to learn to refrain from doing things merely because we know how to do them.
[Editor of the Lancet]
The Lancet 1965 **2** 801

A Frankland 1825–1899

50   I am convinced that the future progress of chemistry as an exact science depends very much upon the alliance with mathematics.
Am. J. Math. 1878 **1** 349

Benjamin Franklin 1706–1790

51   He snatched lightning from the heavens and sceptres from kings.
[Epitaph on him by Turgot]

52   To study, to finish, to publish.

[Sir] James George Frazer 1854–1941

53   But while science has this much in common with magic that both rest on a faith in order as the underlying principle of all things ... magic differs widely from that which forms the basis of science.
The Golden Bough abridged edn, 1925 (London: Macmillan)

Frederick the Great [King of Prussia] 1712–1786

54   I have no fault to find with those who teach geometry. That science is the only one which has not produced sects; it is founded on analysis and on synthesis and on the calculus; it does not occupy itself with probable truth; moreover it has the same method in every country.
Oeuvres ed Decker, 7, p 100

Gottlob Frege 1848–1925

55   A scientist can hardly meet with anything more undesirable than to have the foundation give way just as the work is finished. I was put in this position by a letter from Mr Bertrand Russell when the work was nearly through the press.
In *Scientific American* May 1984, p 77

Augustin Jean Fresnel 1788–1827

56   Nature is not embarrassed by difficulties of analysis.

57   If you cannot saw with a file or file with a saw, then you will be no good
      as an experimentalist.
      In C V Boys *DSB*

Sigmund Freud 1856–1939

58   I am not really a man of science, not an observer, nor an experimenter,
      and not a thinker. I am by temperament nothing but a conquistador ...
      with the curiosity, the boldness and the tenacity that belong to that type
      of person.
      In E Jones *Life and Work of Sigmund Freud* 1953 (London: Hogarth) vol 1, p 348

59   I have no concern with any economic criticisms of the communistic system:
      I cannot inquire into whether the abolition of private property is advan-
      tageous and expedient. But I am able to recognize that psychologically it
      is founded on an untenable illusion. By abolishing private property one
      deprives the human love of agression of one of its instruments .... This
      instinct did not arise as the result of property; it reigned almost supreme
      in primitive times when possessions were still extremely scanty.
      An Outline of Psychoanalysis (last postumous essays). In R Ardey *The Territorial
      Imperative* 1967 (London: Collins) p 293

60   My life and work has been aimed at one goal only: to infer or guess how the
      mental apparatus is constucted and what forces interplay and counteract
      in it.
      See E Jones *Life and Work of Sigmund Freud* 1953 (London: Hogarth) vol 1

61   [Poets] are masters of us ordinary men, in knowledge of the mind, because
      they drink at streams which we have not yet made accessible to science.

Max J Friedländer

62   Art being a thing of the mind, it follows that any scientific study of art
      will be psychology. It may be other things as well, but psychology it will
      always be.
      Von Kunst und Kennerschaft, Oxford and Zürich, 1946, p 128

Robert Frost 1874–1963

63   Did I see it go by,
      That Millikan mote?
      Well, I said that I did.
      I made a good try.
      But I'm no one to quote.
      If I have a defect
      It's a wish to comply
      And see as I'm bid.
      I rather suspect
      All I saw was the lid
      Going over my eye,
      I honestly think
      All I saw was a wink.
      [On being asked to look at Millikan's oil-drop experiment for determining the charge
      on the electron]
      Complete Poems of Robert Frost 1951 (London: Cape) p 424

64   Some say the world will end in fire,
      Some say in ice.
      From what I've tasted of desire
      I hold with those who favour fire.
      The Poetry of Robert Frost 1969 (London: Cape) p 220

65   The telescope at one end of his beat,
      And at the other end the microscope,
      Two instruments of nearly equal hope, ...
      The Bear in *The Poetry of Robert Frost* 1969 (London: Cape) p 269

66   We dance round in a ring and suppose,
      But the Secret sits in the middle and knows
      The Poetry of Robert Frost (London: Cape) p 362

Richard Buckminster Fuller 1895–1983

67   I am a passenger on the spaceship, Earth.
      Operating Manual for Spaceship Earth 1969 (New York: Pocket Books)

68   How much does your building weigh?

Roy Fuller 1912–

69   Poetry as our culture knows it ... is ... over-determined.
      Professors and Gods 1973 (London: Deutsch) p 105

70   There is something almost intrinsically poetic in the scientific.
      Professors and Gods 1973 (London: Deutsch) p 53

Dennis Gabor 1900–1979

1   The most important and urgent problems of the technology of today are
no longer the satisfactions of the primary needs or of archetypal wishes,
but the reparation of the evils and damages wrought by the technology of
yesterday.
Innovations: Scientific, Technological and Social 1970 (Oxford: Oxford University
Press) p 9

2   Short of a compulsory humanistic indoctrination of all scientists and engineers, with a 'Hippocratic oath' of never using their brains to kill people,
I believe that the best makeshift solution at present is to give the alpha-
minuses alternative outlets for their dangerous brain-power, and this may
well be provided by space research.
Inventing the Future 1964 (Harmondsworth: Penguin) p 135

3   Till now man has been up against Nature; from now on he will be up against
his own nature.
Inventing the Future 1964 (Harmondsworth: Penguin) p 89

4   The future cannot be predicted, but futures can be invented. It was man's
ability to invent which has made human society what it is. The first step
of an inventor is to visualise, by an act of imagination, a thing or state
which does not yet exist and which to him appears in some way desirable.
He can then start rationally arguing backwards and forwards until a way
is found from one to the other. For the social inventor the engineering of
human consent is the most essential and the most difficult step.
Obituary in *Biogr. Mem. FRS* **26** 125 (1980)

John Kenneth Galbraith 1908–

5   The real accomplishment of modern science and technology consists in taking ordinary men, informing them narrowly and deeply and then, through
appropriate organisation, arranging to have their knowledge combined with
that of other specialised but equally ordinary men. This dispenses with the
need for genius. The resulting performance, though less inspiring, is far
more predictable.
The New Industrial State 1969 (Harmondsworth: Penguin) p 71

6   [On Mrs Margaret H Thatcher] She is a reflection of comfortable middle-
class values that do not take seriously the continuing unemployment. What
I particularly regret is that she does not take seriously the intellectual
decline. Having given up the Empire and the mass production of industrial
goods, Britain's future lay in its scientific and artistic pre-eminence. Mrs
Thatcher will be long remembered for the damage she has done.
The Guardian 15 October 1988 p 21

7   It was with Malthus and Ricardo that economics became the dismal science.
The Age of Uncertainty

8   The enemy of the market is not the ideology but the engineer.
The New Industrial State 1967

9   There is a further case for reality as opposed to institutional truth. That
is the very considerable personal pleasure to be found in pursuing it. To
the adherents of the institutional truth there is nothing more inconvenient,

nothing that so contributes to discomfort, than open, persistent articulate
assertion of what is real.
Commencement Address to Smith College, Massachusetts, *The Guardian* 28 July 1989
p 23

[Abbé] Célestin Galiani 1681–1753

10   *Les dés de la nature sont pipés.*
The dice of nature are loaded.
On Growth and Form D'Arcy Thomson

Galileo Galilei 1564–1642

11   I, Galileo Galilei, son of the late Vincenzio Galilei of Florence, aged seventy years, being brought personally to judgment, and kneeling before you, Most Eminent and Most Reverend Lords Cardinals, General Inquisitors of the Universal Christian Commonwealth against heretical depravity, having before my eyes the Holy Gospels which I touch with my own hands, swear that I have always believed, and, with the help of God, will in future believe, every article which the Holy Catholic and Apostolic Church of Rome holds, teaches, and preaches. But because I have been enjoined, by this Holy Office, altogether to abandon the false opinion which maintains that the Sun is the centre and immovable, and forbidden to hold, defend, or teach, the said false doctrine in any manner...I am willing to remove from the minds of your Eminences, and of every Catholic Christian, this vehement suspicion rightly entertained towards me, therefore, with a sincere heart and unfeigned faith, I abjure, curse, and detest the said errors and heresies, and generally every other error and sect contrary to the said Holy Church; and I swear that I will never more in future say, or assert anything, verbally or in writing, which may give rise to a similar suspicion of me; but that if I shall know any heretic, or anyone suspected of heresy, I will denounce him to this Holy Office, or the the Inquisitor and Ordinary of the place in which I may be. I swear, moreover, and promise that I will fulfil and observe fully all the penances which have been or shall be laid on me by this Holy Office. But if it shall happen that I violate any of my said promises, oaths, and protestations (which God avert), I subject myself to all the pains and punishments which have been decreed and promulgated by the sacred canons and other general and particular constitutions against delinquents of this description. So, may God help me, and his Holy Gospels, which I touch with my own hands, I, the above named Galileo Galilei, have abjured, sworn, promised, and bound myself as above; and, in witness thereof, with my own hand have subscribed this present writing of my abjuration, which I have recited word for word.
In J J Fahie *Galileo, His Life and Work* transl I B Cohen, 1903 p 313

12   In questions of science the authority of a thousand is not worth the humble reasoning of a single individual.
Arago's Eulogy of Laplace Smithsonian Report, 1874 p 164

13   Philosophy is written in that great book which ever lies before our gaze—I mean the universe—but we cannot understand if we do not first learn the language and grasp the symbols in which it is written. The book is written in the mathematical language, and the symbols are triangles, circles and other geometrical figures, without the help of which it is impossible to

conceive a single word of it, and without which one wanders in vain through a dark labyrinth.
Opere Il Saggiatore p 171

14 Take note, theologians, that in your desire to make matters of faith out of propositions relating to the fixity of sun and earth you run the risk of eventually having to condemn as heretics those who would declare the earth to stand still and the sun to change position—eventually, I say, at such a time as it might be physically or logically proved that the earth moves and the sun stands still.
Dialogue

15 *Epur si muove*
And yet it does move
Aprocryphal words to himself after making his abjuration of heliocentricity

16 I wish, my dear Kepler, that we could have a good laugh together at the extraordinary stupidity of the mob. What do you think of the foremost philosophers of this University? In spite of my oft-repeated efforts and invitations, they have refused, with the obstinacy of a glutted adder, to look at the planets or the Moon or my glass [telescope].
[Through which the satellites of Jupiter were visible—seen first in January 1610.]
Opere ed Naz. X p423

17 The knowledge of a single fact acquired through a discovery of its causes, prepares the mind to understand and ascertain other facts, without need of recourse to experiments, precisely as in the present case, where by argumentation alone the author proves with certainty that the maximum range occurs when the elevation is forty-five degrees.
In S F Mason *A History of the Sciences* 1953 (London: Routledge) p 123

Francis Galton 1822–1911

18 Whenever you can, count.
In *The World of Mathematics* ed J R Newman 1956 (New York: Simon & Schuster) p 1169

19 [statistics are] the only tools by which an opening can be cut through the formidable thicket of difficulties that bars the path of those who pursue the Science of Man.
In Karl Pearson *The Life, Letters and Labours of Francis Galton* 1914 (London: Cambridge University Press)

20 The furniture of a man's mind chiefly consists of his recollections and the bonds which unite them.
1883

21 The scientific advantages of travel are enormous to a man prepared to profit by them. He sees Nature working by herself, without the interference of human intelligence; and he sees her from a new point of view; he

has also undisturbed leisure for the problems which perpetually abstract his attention by their novelty. The consequence is, that though scientific travellers are comparatively few, yet out of their ranks a large proportion of the leaders in all branches of science has been supplied.
The Art of Travel 4th edn, 1867 (London: John Murray) p 2

22   [Letter to Darwin, 24 December 1869] I always think of you in the same way as converts from barbarism think of the teacher who first relieved them from the intolerable burden of their superstitions...Consequently the appearance of your *Origin of Species* formed a real crisis in my life; your book drove away the constraint of my old superstition as if it had been a nightmare, and was the first to give me freedom of thought.
In E R Trattner *Architects of Ideas* 1938 (New York: Carrick and Evans)

Mohandas Karamchand Gandhi 1869–1948

23   [Asked, on his arrival in Europe, what he thought of Western civilization]: 'I think it would be an excellent idea'.
[Attributed]

24   I do not remember to have seen a handloom or a spinning wheel when in 1908 I described it in Hind Swaraj as the panacea for the growing pauperism of India.
Autobiography 1927 1982 (Harmondsworth: Penguin) p 439

Martin Gardner 1914–

25   Any magician will tell you that scientists are the easiest persons in the world to fool...In their laboratories the equipment is just what it seems. There are no hidden mirrors or secret compartments or concealed magnets...The thinking of a scientist is rational, based on a lifetime of experience with a rational world. But the methods of magic are irrational and totally outside a scientist's experience.
Bul. Sci., Bangalore March–April 1989 p 41

Charles de Gaulle 1890–1970

26   No country without an atomic bomb could properly consider itself independent.
New York Times Magazine 12 May 1968

Karl Friedrich Gauss 1777–1855

27   ...*durch planmässiges Tattonieren.*
...through systematic, palpable experimentation.
[Asked how he came upon his theorems]

28   God does arithmetic.
[Attributed]

29   I have had my results for a long time: but I do not yet know how I am to arrive at them.
In A Arber *The Mind and the Eye* 1954 p 47

30   We must admit with humility that, while number is purely a product of our minds, space has a reality outside our minds, so that we cannot completely prescribe its properties *a priori.*
Letter to Bessel, 1830

George III 1738–1820

31   I spend money on war because it is necessary, but to spend it on science, that is pleasant to me.
[to Lalande]
In R A Gregory *Discovery...* 1918 (London: Macmillan) p 47

Hans Heinrich Gerth and Charles Wright Mills 1908– and 1916–1962

32   Precisely because of their specialization and knowledge, the scientist and technician are among the most easily used and coordinated of groups in modern society...the very rigor of their training typically makes them the easy dupes of men wise in political ways.
Character & Social Structure 1954 (London: Routledge & Kegan Paul)

Al-Ghazali 1058–1111

33   There is no hope in returning to a traditional faith after it has once been abandoned, since the essential condition in the holder of a traditional faith is that he should not know that he is a traditionalist.
In E R Dodds *The Greeks and the Irrational* 1951 (Berkeley, CA: University of California Press)

Edward Gibbon 1737–1794

34   Twenty-two acknowledged concubines, and a library of sixty-two thousand volumes, attested the variety of his inclinations; and from the productions which he left behind him, it appears that the former as well as the latter were designed for use rather than for ostentation. (By each of his concubines, the younger Gordian left three or four children. His literary productions, though less numerous, were by no means contemptible.)
[Roman Emperor, died 238]
The Decline and Fall of the Roman Empire Chapter 7

35   The winds and waves are always on the side of the ablest of navigators.
The Decline and Fall of the Roman Empire

Josiah Willard Gibbs 1839–1903

36   One of the principal objects of theoretical research in my department of knowledge is to find the point of view from which the subject appears in its greatest simplicity.
1881

37   Mathematics *is* a language.
At a Yale faculty meeting

William Gilbert 1540–1603

38   Look for knowledge not in books but in things themselves.
De Magnete

J Gillis

39   [Weizmann Institute] It is not generally realised outside of academic circles how far a mediocre research worker can get. ...And without this great intellectual proletariat of research how far should we get?
In Aldous Huxley *Literature and Science* 1963 p 62–3

Allen Ginsberg 1926–

40   The war is language,
      language abused
         for Advertisement
      language used
      like magic for power on the planet
      Black Magic language
      formulas for reality–
      Communism is a 9 letter word
         used by inferior magicians
      with the wrong alchemical formula for transforming
         earth into gold
      funky warlocks operating on guesswork,
      hand-me-down mandrake terminology.
      Wichita Vortex Sutra in *The East-side Scene: American Poetry 1960–65* ed A De
      Loach 1972 (NY: Dobleday)

41   I saw the best minds of my generation, destroyed by madness, starving,
      hysterical, naked.
      First line of *Howl* ca 1956

Thomas Favill Gladwin 1917–

42   No style of thinking will survive which cannot produce a usable product
      when survival is at stake.
      [On the navigation of the Puluwat Islanders]
      East is a Big Bird: Navigation and Logic on Puluwat Atoll 1970 (Cambridge, MA:
      Harvard University Press)

Max Gluckman 1911–

43   A science is any discipline in which the fool of this generation can go beyond
      the point reached by the genius of the last geneation.
      Politics, Law and Ritual 1965 (NY: Mentor) p 60

Robert H Goddard 1882–1945

44   [US rocket pioneer] God pity a one-dream man.
      Notebooks, Goddard Memorial Library, Clark University, USA. In C Sagan *Broca's
      Brain* 1980 New York: Coronet p 281

Kurt Goedel 1906–1978

45   It is impossible to demonstrate the non-contradictoriness of a logical math-
      ematic system using only the means offered by the system itself.
      [Paraphrased]
      [See E Nagel and J R Newman *Goedel's Proof*]
      Monatshefte für Mathematik und Physik, Leipzig 1931 pp 173–98

Hermann Goering 1893–1946

46   *Wenn ich "Kultur" höre,...entsichere ich meinen Browning.*
      When I hear the word 'culture' I slip the safety-catch of my Browning.
      [Or, attributed: When I hear the word 'culture' I reach for my revolver.]
      [Browning = type of firearm. Also the poet Robert Browning (1812–1889)!]

Johann Wolfgang von Goethe 1749–1832

47   (Laboratorium im Sinne des Mittelalters, weitläufige unbehilfliche

Apparate zu phantastischen Zwecken.)
Wagner: *Es wird ein Mensch gemacht.*
... nun lässt sich wirklich hoffen,
Dass, wenn wir aus viel hundert Stoffen
Durch Mischung—denn auf Mischung kommt es an—
den Menschenstoff gemächlich komponieren,
In einen Kolben verlutieren
Und ihn gehörig kohobieren
So ist das Werk im stillen abgetan.
Es Wird! Die Masse regt sich klarer!
Die Überzeugung wahrer, wahrer:
Was man an der Natur Geheimnisvolles pries,
Das wagen wir verständig zu probieren,
Und was sie sonst organisieren liess
Das lassen wir kristallisieren.
Mephistopheles: *Wer lange lebt, hat viel erfahren,*
Nichts Neues kann für ihn auf dieser Welt geschehn.
Ich habe schon in meinen Wanderjahren
Kristallisiertes Menschenvolk gesehn.
(Laboratory, after the style of the Middle Ages: extensive, unwieldy
apparatus, for fantastical purposes.)
Wagner: A human being in the making...
Look, There's a gleam—
Now hope may be fulfilled,
That hundreds of ingredients, mixed, distilled—
And mixing is the secret—give us power
The stuff of human nature to compound;
If in a limbeck we now seal it round
And cohobate with final care profound,
The finished work may crown this silent hour.
It works. The substance stirs, is turning clearer.
The truth of my conviction presses nearer:
The thing in Nature as high mystery prized,
This has our science probed beyond a doubt;
What Nature by slow process organised,
That have we grasped, and crystallised it out.
Mephistopheles: He who lives long a host of things will know,
The world affords him nothing new to see.
Much have I seen, in wandering to and fro,
Including crystallised humanity.
Faust II, Akt II. Transl Philip Wayne, 1949 (London: Penguin Classics)

48   *'Ins Innre der Natur'*—
O du Philister—
Natur hat weder Kern
Noch Schale.
'In the inside of Nature'—
O you Philistines—
Nature has neither kernel
Nor shell.
Allerdings. Dem Physiker 1819/20

49   'Questions of science,' remarked Goethe, 'are very frequently career questions. A single discovery may make a man famous and lay the foundations
of his fortunes as a citizen...Every newly observed phenomenon is a discovery, every discovery is property. Touch a man's property and his passions
are easily aroused.'
In J P Eckerman *Conversations with Goethe* 21 December 1823

50   *Auf theoretischem Feld ist weiter nichts mehr zu finden; Aber der praktische*
Satz gilt doch; Du kannst, denn du sollst.
In the theoretical field there is no more to be found; but the practical
dictum is still valid; you can, for you ought.
Xenien 1797 (jointly with Schiller)

51   *Der kleine Gott...In jeden Quark begräbt er seine Nase.*
This little God [man] pushes his nose into all kinds of rubbish.
[See *Scientific American* July 1968 for the history of the word 'quark' but it is clear
that the *Scientific American* has taken the quotation out of context and that '*Der
kleine Gott*' is not God but man.]
Faust (Frankfurt: Insel-Verlag) Prologue

52   Faust: *Geschrieben steht: 'Im Anfang war das Wort!'*
Hier stock ich schon! Wer hilft mir weiter fort?
Ich kann das Wort so hoch unmöglich schätzen,
Ich muss es anders übersetzen,
Wenn ich vom Geiste recht erleuchtet bin.
Geschrieben steht: 'Im Anfang war der Sinn.'
Bedenke wohl die erste Zeile,
Dass deine Feder sich nicht übereile!
Ist es der Sinn, der alles wirkt und schafft?
Es sollte stehn: 'Im Anfang war die Kraft!'
Doch, auch indem ich dieses niederschreibe,
Schon warnt mich was, dass ich dabei nicht bleibe.
Mir hilft der Geist! Auf einmal seh ich Rat
Und schreibe getrost: 'Im Anfang war die Tat!'
Faust: Tis writ:'In the beginning was the Word!'
I pause, to wonder what is here inferred?
The Word I cannot set supremely high,
A new translation I will try.
I read, if by the spirit I am taught,
This sense: 'In the beginning was the Thought'.
This opening I need to weigh again,
Or sense may suffer from a hasty pen.
Does Thought create, and work, and rule the hour?
'Twere best: 'In the beginning was the Power!'
Yet, while the pen is urged with willing fingers,
A sense of doubt and hesitancy lingers.
The spirit come to guide me in my need,
I write, 'In the beginning was the Deed!'
Faust I. Transl Philip Wayne, 1949 (London: Penguin Classics)

53   *Doch, der den Augenblick ergreift,*
Das ist der rechte Mann.
He who seizes the right moment,

Is the right man.
Faust I, iii

54 The history of science is science itself; the history of the individual, the individual.
Mineralogy and Geology

55 Nothing is more terrible than to see ignorance in action.
Maxims and Reflections I

56 Thus I saw that most men only care for science so far as they get a living by it, and they worship error when it affords them a subsistence.
In J P Eckerman *Conversations with Goethe* 15 October 1825

57 *Denn eben wo Begriffe fehlen,*
Da stellt ein Wort zur rechten Zeit sich ein.
Mit Worten lässt sich trefflich streiten,
Mit Worten ein System bereiten.
For where ideas fail,
A word at the right time can work wonders for you
And save the situation.
With words you can build a system.
Faust I

58 *Sage Herdern, dass ich dem Geheimnis der Pflanzen-zeugung und -organisation ganz nah bin und dass es das einfachste ist, was nur gedacht werden kann.*
Tell Herder that I am quite near to the secret of the generation and organisation of plants and that it is the simplest that could be conceived of.
Letter to Charlotte von Stein, 8 June 1787

59 Faust: *Das Pentagramma macht dir Pein?*
...Mephistopheles: *Beschaut es recht!*
Es ist nicht gut gezogen;
Der eine Winkel, der nach aussen zu,
Ist, wie du siehst, ein wenig offen.
Faust: You find my pentagram embarrassing?
...*Mephistopheles:* Beg pardon, Sir, the drawing's not completed,
For here, this angle on the outer side
Is left, you notice, open at the joint.
Faust I

60 As for what I may have done as a poet, I take no pride in it whatever...Excellent poets have lived at the same time with me, poets more excellent lived before me, and others will come after me. But that in my century I am the only person who knows the truth in the difficult science of colours—of that, I say, I am not a little proud, and here I have a consciousness of superiority to many.
Conversations with Eckermann

61 *Magnetes Geheimnis, erkläre mir das!*
Kein grösser Geheimnis als Lieb und Hass.
The secret of magnetism, now explain that to me!
There is no greater secret, except love and hate.
Gott, Gemüt und Welt

62   Gray and ashen, my friend, is every science. And only the golden tree of life is green.
Faust

63   *Dauer im Wechsel.*
Duration in change.
[Favourite expression]

64   *Dann im Kristall und seiner ewigen Schweignis*
Erblicken sie der Oberwelt Ereignis.
In the eternal silence within a crystal they may see the happenings of the world outside.
Faust II, Act 4, *Auf dem Vorgebirg.*

Thomas Gold 1920–

65   Things are as they are because they were as they were.
Misner, Thorpe and Wheeler *Gravitation* (San Francisco: Freeman)

Maurice Goldsmith

66   Bernal and Joliot-Curie believed passionately in the social function of science for solving the problems of mankind but, as with the early pioneers of X-rays and radium, over-exposure to the matter of their studies contributed greatly to their deaths.

Brian Carey Goodwin 1931–

67   The discovery of appropriate variables for biology is itself an act of creation.
In *Towards a Theoretical Biology* 2 ed C H Waddington, 1969 (Edinburgh: Edinburgh University Press)

George Gore 1826–1908

68   A geometrical and mechanical basis of physical science cannot be constructed until we know the forms, sizes, and positions of the molecules of substances.
The Art of Scientific Discovery or the General Conditions and Methods of Research in Physics and Chemistry 1878 (London) pp 26–29

Stephen Jay Gould 1941–

69   A man does not attain the status of Galileo merely because he is persecuted; he must also be right.
Ever since Darwin 1977 (New York: Norton) p 154

70   Science is all those things which are confirmed to such a degree that it would be unreasonable to withold one's provisional consent.
Lecture on Evolution, Cambridge, 1984

M H Goyns

71   I would suggest that British science is as alive and well as it has ever been. It is just that it is currently living in the United States.
Nature **344** 98, 8 March 1990

Walter Gratzer

72   The common perception of science as a rational activity, in which one confronts the evidence of fact with an open mind, could not be more false.

Facts assume significance only within a pre-existing intellectual structure, which may be based as much on intuition and prejudice as on reason.
The Guardian 28 September 1989

Robert Ranke Graves 1895–1985

73 'The sum of all the parts of Such—
Of each laboratory scene—
Is such.' While Science means this much
And means no more, why let it mean!
    But were the science-men to find
Some animating principle
Which gave synthetic Such a mind
Vital, though metaphysical—
To Such, such an event, I think
Would cause unscientific pain:
Science, appalled by thought, would shrink
To its component parts again.
Poems 1926–1930 1931 (London: Heinemann)
Synthetic Such 1966 (Harmondsworth: Penguin) p 79

74 Myth, then, is a dramatic shorthand record of such matters as invasions, migrations, dynastic changes, admissions of foreign cults, and social reforms.
Introduction to the Larousse Encyclopedia of Mythology 1959 (London: Hamlyn)

75 Thought comes often clad in the strangest clothing:
So Kekulé the chemist watched the weird rout
Of eager atom-serpents writhing in and out
And waltzing tail to mouth. In that absurd guise
Appeared benzene and aniline, their drugs and their dyes.
Difficult Questions, Easy Answers 1972 (London: Cassell)

76 To know only one thing well is to have a barbaric mind: civilization implies the graceful relation of all varieties of experience to a central humane system of thought. The present age is peculiarly barbaric: introduce, say, a Hebrew scholar to an ichthyologist or an authority on Danish place names and the pair of them would have no single topic in common but the weather or the war (if there happened to be a war in progress, which is usual in this barbaric age).
Saturday Review **46** 82–8, 7 December 1963

77 Is it not the height of silent humour
To cause an unknown change in the earth's climate?
The Meeting. In C Sagan's *Broca's Brain* p 235

Robin Gray 1942–

78 [British surgeon in the Red Cross Hospital in Peshawar] There's an appalling number of paraplegics. It's a particular feature of modern weaponry. It always amazes me that there are people calling themselves human beings who spend their lives designing weapons which maximise soft tissue injury.
The Guardian 12 November 1987

Thomas Gray 1716–1771

79 Alas! regardless of their doom,

The little victims play!
No sense have they of ills to come,
Nor care beyond today.
Ode on a Distant Prospect of Eton College

80   Fair science frown'd not on his humble birth,
And melancholy mark'd him for her own.
Elegy written in a Country Churchyard

[Sir] Richard Arman Gregory 1864–1952

81   Science is not to be regarded merely as a storehouse of facts to be used for
material purposes, but as one of the great human endeavours to be ranked
with arts and religion as the guide and expression of man's fearless quest
for truth.

Richard Langton Gregory 1923–

82   On how so little information controls so much behaviour.
In *Towards a Theoretical Biology* 2 ed C H Waddington, 1969 (Edinburgh: Edinburgh
University Press) p 236

Christopher Murray Grieve [Hugh McDiarmid] 1892–1978

83   Perchance the best chance of reproducing the ancient Greek temperament
would be to cross the Scots with the Chinese.
Lucky Poet 1943 (London: Methuen)

Juan Gris 1887–1927

84   When it began, cubism was a sort of analysis, which was no more painting
than the description of physical phenomena was physics.
In D Kahnweiler *Juan Gris, his Life and Work* 1947 (New York: Valentine)
In *Physics Bulletin* February 1975 p 61

85   It is not picture X which manages to correspond with my subject, but the
subject X which manages to correspond with my picture...The mathematics of picture-making lead me to the physics of representation.
In Herbert Read *Icon and Idea* 1955 (London: Faber) p 130

[General] Leslie Richard Groves 1896–1970

86   [Officer in command of the US Atomic Bomb Installations] Compartmen-
talization of knowledge, to me, was the very heart of security. My rule
was simple and not capable of misinterpretation—each man should know
everything he needed to know to do his job and nothing else.
[Almost the classic recipe for preventing originality]
Now it can be told 1962 (New York: Harper & Row) p 140

Branko Grünbaum 1926– and G C Shephard

87   Mathematicians have long since regarded it as demeaning to work on problems related to elementary geometry in two or three dimensions, in spite
of the fact that it is precisely this sort of mathematics which is of practical
value.
Handbook of Applicable Mathematics ed W Ledermann and S Vajda 1985 (London:
Wiley) vol VB p 728

Ernesto [Che] Guevara 1928–1967

88   When asked whether or not we are Marxists, our position is the same as
      that of a physicist or a biologist who is asked if he is a 'Newtonian', or if
      he is a 'Pasteurian'.
      In *Radical Currents in Contemporary Philosophy* ed David DeGrood 1971 (St Louis,
      MO: Warren Green) p 175

Dennis Gunton

89   [British Council Representative in Yugoslavia]
      Gunton's Laws of Librarianship
      11th Law: The value of rubbish is unaltered by translation, abstraction or
      citation.
      13th Law: Anyone who can be replaced by a machine deserves to be.
      Libr. Ass. Rec. 1990 **92** (1) 48–9

Fritz Haber 1868–1934

1   For more than forty years I have selected my collaborators on the basis of
their intelligence and their character and I am not willing for the rest of
my life to change this method which I have found so good.
Letter of Resignation 30 April 1933
[Haber was unwilling to follow the Nazi requirements for racial purity.]

2   [Originator of the first use of chlorine gas on 22 April 1915 at Ypres.]
In no future war will the military be able to ignore poison gas. It is a higher
form of killing.
Nobel Prize acceptance speech, 1919. *Nature* 8 July 1982 **298** 205

Hadrian [Publius Aelius Hadrianus][Emperor 117–139] 76–138

3   *Animula vagula, blandula*
Hospes comesque corporis,
Quae nunc abibis in loca
Pallidula, rigida nudula;
Nec, ut soles, dabis jocos?
Dying address to his soul

Ernst Heinrich Haeckel 1834–1919

4   God...[is]...a gaseous vertebrate.
The riddle of the Universe 1929 (London: Watts) p 235

5   Ontogeny recapitulates phylogeny. [In full] Ontogenesis, or the development
of the individual is a short and quick recapitulation of phylogenesis, of the
development of the tribe to which it belongs, determined by the laws of
inheritance and adaptation.
The History of Creation 1868
[See S J Gould *Ontogeny and Phylogeny* 1977 (Harvard: Belknapp)]

6   *Krystallographie der Organismen.*
A crystallography of organisms.
Generelle Morphologie der Organismen vol I 1866 (Berlin: Georg Reimer) p 139

7   *Der Mensch schafft Gott nach seinem Bilde.*
Man creates God in his own image.
Generelle Morphologie vol I 1866 (Berlin: Georg Reimer) p 174

[Lord] Hailsham [Quintin Hogg] 1907–

8   None the less I am, so far as I know, the first, and possibly the only Minister
for Science (or of Science for that matter) in the Universe...
[There was at least one other—in India]
Science and Politics 1963 (London: Faber & Faber) p 9

John Burdon Sanderson Haldane 1892–1964

9   Cancer's a Funny Thing:
I wish I had the voice of Homer
To sing of rectal carcinoma,
Which kills a lot more chaps, in fact,
Than were bumped off when Troy was sacked...
[Written while mortally ill with cancer]
In Ronald Clark *JBS* 1968 (London: Hodder & Stoughton) p 257

10   The conservative has but little to fear from the man whose reason is the servant of his passions, but let him beware of him in whom reason has become the greatest and most terrible of the passions.
Daedalus, or Science and the Future 1923 (London: Kegan Paul) p 78

11   I have no doubt that in reality the future will be vastly more suprising than anything I can imagine. Now my own suspicion is that the universe is not only queerer than we suppose, but queerer than we can suppose.
Possible Worlds and Other Papers 1927 (London: Chatto and Windus) p 286

12   I'd lay down my life for two brothers or eight cousins.
New Scientist 8 August 1974 p 325

13   In scientific thought we adopt the simplest theory which will explain all the facts under consideration and enable us to predict new facts of the same kind. The catch in this criterion lies in the word 'simplest'. It is really an aesthetic canon such as we find implicit in our criticisms of poetry or painting. The layman finds such a law as $dx/dt = K(d^2x/dy^2)$ much less simple than 'it oozes', of which it is the mathematical statement. The physicist reverses this judgment, and his statement is certainly the more fruitful of the two, so far as prediction is concerned. It is, however, a statement about something very unfamiliar to the plain man, namely, the rate of change of a rate of change.
Science and theology as Art Forms in *Possible Worlds* 1927 (London: Chatto and Windus) p 227

14   Religion is a way of life and an attitude to the universe. It brings man into closer touch with the inner nature of reality. Statements of fact made in its name are untrue in detail, but often contain some truth at their core. Science is also a way of life and an attitude to the universe. It is concerned with everything but the nature of reality. Statements of fact made in its name are generally right in detail, but can only reveal the form and not the real nature of existence. The wise man regulates his conduct by the theories both of religion and science. But he regards these theories not as statements of ultimate fact, but as art forms.
Science and theology as Art Forms in *Possible Worlds* 1927 (London: Chatto and Windus) p 239

15   Religion is still parasitic in the interstices of our knowledge which have not yet been filled. Like bed-bugs in the cracks of walls and furniture, miracles lurk in the lacunae of science. The scientist plasters up these cracks in our knowledge; the more militant Rationalist swats the bugs in the open. Both have their proper sphere and they should realise that they are allies.
Science and Life: Essays of a Rationalist 1968 (London: Pemberton and Barrie & Rockliff)
In *The Rationalist Annual* 1934

16   A time will however come (as I believe) when physiology will invade and destroy mathematical physics, as the latter has destroyed geometry.
Daedalus, or Science and the Future 1923 (London: Kegan Paul) p 15

17   We are part of history ourselves, and we cannot avoid the consequences of being unable to think impartially.
Heredity and Politics 1938 (London: Allen & Unwin) p 126

18   Why cannot people learn to speak the truth? I have, I think, taught two, perhaps three, Indian colleagues to do so. It will probably wreck their careers.
In Ronald Clark *JBS* 1968 (London: Hodder & Stoughton) p 262

19   1. Write down the structural formula of human type C oxyhaemoglobin, and briefly summarize the evidence on which it is based. (Structural formulae should be written stereoscopically. A stereoscope is provided.)
Question in a mock exam paper looking forward 25 years from 1931. *Brighter Biochemistry* 1931. Reprinted TIBS, March 1981 p 91

20   He had greatly enjoyed this dinner, 'thanks to the delectable company of the young lady on my right and the young lady on my left. Now in a three-dimensional world one can have only two ladies sitting next to one. Heaven, I believe, might be conceived to be, a place in N-dimensional space, where one could therefore expect at dinner to enjoy the company of $N - 1$ young ladies'.
After-dinner speech. In E Ashby *Nature* 28 April 1977 **266** 782

Stephen Hales 1677–1761

21   Since we are assured that the all-wise Creator has observed the most exact proportions, of number, weight and measure, in the make of all things, the most likely way therefore, to get any insight into the nature of those parts of the creation, which come within our observation, must in all reason be to number, weigh and measure.
Vegetable Staticks Introduction
[Wisdom of Solomon 11:20]

John Hall 17th Century

22   If that this` thing we call the world
By chance on atoms was begot
Which though in ceaseless motion whirled
Yet weary not
How doth it prove
Thou art so fair and I in love.
Epicurean Ode

[Sir] William Rowan Hamilton 1805–1865

23   Here, as he walked by on the 16th October, 1843, Sir William Rowan Hamilton in a flash of genius discovered the fundamental formula for quaternion algebra $[i^2 = j^2 = k^2 = ijk = -1]$ and cut it on a stone of this bridge.
[Engraved on a stone of Brougham Bridge, over the Royal Canal, Dublin]
[See John Milton *Paradise Lost* v, 181: 'ye Elements...that in quaternion run'.

Norwood Russell Hanson 1924–1967

24   There is more to seeing than meets the eyeball.
Patterns of Discovery 1958 (Cambridge: Cambridge University Press)

E Y Harburg

25   [Librettist for the 'Wizard of Oz']
Leave de atom alone.
...Don't get smart alecksy

Wid de galaxy
Leave de atom alone.
(1957)
London Review of Books 1984 **6**, No 1, p 4

Godfrey Harold Hardy 1877–1947

26   ...a science is said to be useful if its development tends to accentuate the
existing inequalities in the distribution of wealth, or more directly promotes
the destruction of human life.
A Mathematician's Apology 1941 (London: Cambridge University Press)

27   Beauty is the first test; there is no permanent place in the world for ugly
mathematics.
A Mathematician's Apology 1941 (London: Cambridge University Press)

28   I have never done anything 'useful'. No discovery of mine has made, or is
likely to make, directly or indirectly, for good or ill, the least difference to
the amenity of the world. ... Judged by all practical standards, the value of
my mathematical life is nil; and outside mathematics it is trivial anyhow. I
have just one chance of escaping a verdict of complete triviality, that I may
be judged to have created something is undeniable; the question is about
its value.
A Mathematician's Apology 1941 (London: Cambridge University Press)

29   There is no scorn more profound, or on the whole more justifiable, than
that of the men who make for the men who explain. Exposition, criticism,
appreciation, is work for second-rate minds.
A Mathematician's Apology 1941 (London: Cambridge University Press)

30   *Reductio ad absurdum*...is a far finer gambit than any chess gambit: a chess
player may offer the sacrifice of a pawn or even a piece, but a mathematician
offers the game.

31   [Hardy said that if you can prove two contradictory theorems then you can
prove anything. He was then challenged to prove, given that $2 + 2 = 5$,
that McTaggart is the Pope.] We also know that $2 + 2 = 4$, so that $5 =
4$. Subtracting 3 we get $2 = 1$. McTaggart and the Pope are two, hence
McTaggart and the Pope are one.
In I Stewart *Concepts of Modern Mathematics* 1975 (Harmondsworth: Penguin) p 116

32   [The word 'intellectual' as used by literary intellectuals in the thirties
seemed not to include Rutherford or Eddington or Dirac or himself.] It
does seem rather odd, don't you know.
C P Snow

[Sir] William Bate Hardy 1864–1934

33   [To Sir Henry Tizard] You know, this applied science is just as interesting
as pure science, and what's more it's a damn sight more difficult.
In Sir Henry Tizard *Haldane Memorial Lecture* Birkbeck College, University of London, 1955

Herbert Amory Hare 1862–1931

34   At first it is impossible for the novice to cast aside the minor symptoms,
which the patient emphasises as his major ones, and to perceive clearly

that one or two facts that have been belittled in the narration of the story of the illness are in reality the stalk about which everything in the case must be made to cluster.

Practical Diagnosis 1899 (Philadelphia, PÀ: Lea Bros)

S J Harris

35   The real danger is not that computers will begin to think like men, but that men will begin to think like computers.

[Sir] Basil Henry Liddell Hart 1895–

36   Surprise is the supreme virtue of warfare, originality of mind the quality that breeds it.

William Harvey 1573–1657

37   *Ex ovo omnia.*
      Everything from an egg.
      De Generatione Animalium London, 1651. Frontispiece. See I B Cohen in *Changing Perspectives in the History of Science. Essays in Honour of Joseph Needham* ed M Teich and R Young 1973 (London: Heinemann) pp 233–249

Herbert George [Blondie] Hasler 1914–

38   You cannot have the success without the failures.
      [Organiser of the single-handed Atlantic yacht race]
      The Observer 7 July 1968

Stephen William Hawking 1942–

39   God not only plays dice. He also sometimes throws the dice where they cannot be seen.
      [See Albert Einstein]
      Nature 1975 **257** 362

Friedrich August von Hayek 1899–

40   There are no better terms available to describe this difference between the approach of the natural and the social sciences than to call the former 'objective' and the latter 'subjective'.
      The Counter-revolution of Science 1952 (New York: Free Press) p 28

Oliver Heaviside 1850–1925

41   [Criticised for using formal mathematical manipulations, without understanding how they worked] Should I refuse a good dinner simply because I do not understand the processes of digestion ?
      [The inventor of the operational calculus and predictor of the Cherenkov effect. Now at last his reputation is increasing to a juster estimate]

Georg Friedrich Hegel 1770–1831

42   The soul takes refuge in the realms of thought, and in opposition to the real world it creates a world of ideas. Philosophy begins with the decline of the real world: when she appears with her abstractions, painting grey on grey, then the freshness of youth and life is already gone, and her reconciliation is not one in reality but in an ideal world.
      In S F Mason *A History of the Sciences* 1953 (London: Routledge) p 469

Piet Hein 1905–

43   Nature, it seems, is the popular name
For milliards and milliards and milliards
Of particles playing their infinite game
Of billiards and billiards and billiards.
Atomyriades in *Grooks* 1966 (Cambridge, MA: Harvard University Press)

Ernst Heinkel 1888–1958

44   [Of Dr Ferdinand Porsche (1875–1951), the automobile engineer, designer
of the Volkswagen, the Tiger tank and the Porsche] He is a very amiable
man but let me give you this advice. You must shut him up in a cage with
seven locks and let him design his engine inside it. Let him hand you the
blueprints through the bars. But for heaven's sake don't ever let him see
the drawing or the engine again. Otherwise he'll ruin you.
[Modifications to engineering designs are what cost the money]
He 1000 1956 (London: Hutchinson) p 63

Werner Heisenberg 1901–1976

45   Natural science does not simply describe and explain nature; it is part of
the interplay between nature and ourselves; it describes nature as exposed
to our method of questioning.
Physics and Philosophy 1959 (London: Allen & Unwin)

46   Science clears the fields on which technology can build.

47   An expert is someone who knows some of the worst mistakes that can be
made in his subject, and how to avoid them.
Physics and Beyond ed R N Ashen 1971 (New York: Harper & Row)

48   In the course of coming into contact with empirical material, physicists have
gradually learned how to pose a question properly. Now, proper questioning
often means that one is more than half the way towards solving the problem.
Physik und Philosophie

49   The atom of modern physics can be symbolised only through a partial differential equation in an abstract space of many dimensions. All its qualities
are inferential; no material properties can be directly attributed to it. That
is to say, any picture of the atom that our imagination is able to invent
is for that very reason defective. An understanding of the atomic world in
that primary sensuous fashion...is impossible.
1945

50   The Bavarian unites the discipline of the Austrian with the charm of the
Prussian.
Obituary by E Teller *Nature* 15 April 1976 **260** 658

Hermann Ludwig Ferdinand von Helmholtz 1821–1894

51   Whoever, in the pursuit of science, seeks after immediate practical utility
may rest assured that he seeks in vain.
Academic Discourse Heidelberg, 1862

52   The first discovery of a new law, is the discovery of a similarity which has
hitherto been concealed in the course of natural processes. It is a manifestation of that which our forefathers in a serious sense described as 'wit', it

is of the same quality as the highest performances of artistic perception in the discovery of new types of expression.

On Thought in Medicine in *Popular Lectures on Scientific Subjects* 1884 (London: Longmans) p 227

Claude-Adrien Helvetius 1715–1771

53   *J'ai cru qu'on devait traiter la morale comme toutes les autres sciences, et faire une morale comme une physique expérimentale.*

I have thought that one ought to treat morals as one treats the other sciences and regard morals as experimental physics.

De l'Ésprit 1758 Preface

L J Henderson

54   Science owes more to the steam engine than the steam engine owes to science.

(1917)

See entry under James Bryant Conant

Heraclitus [of Ephesus] ca 550–475 BC

55   We must know that war is common to all and strife is justice, and that all things come into being and pass away through strife.

In G S Kirk *Heraclitus, The Cosmic Fragments* 1954 (London: Cambridge University Press)

56   All things are an exchange for fire, and fire for all things, even as wares for gold and gold for wares.

In S F Mason *A History of the Sciences* 1953 (London: Routledge & Kegan Paul) p 16

57   If you do not expect the unexpected, you will not find it; for it is hard to be sought out, and difficult.

In C H Khan *The Art and Thought of Heraclitus* 1979 (Cambridge: Cambridge University Press)

58   Incredibility escapes recognition.

In C H Kahn *The Art and Thought of Heraclitus* 1979 (Cambridge: Cambridge University Press) LXXXVI

59   The fairest order in the world is a heap of random sweepings.

In C H Kahn *The Art and Thought of Heraclitus* 1979 (Cambridge: Cambridge University Press) CXXV

60   Pythagoras son of Mnesarchus pursued inquiry further than all other men and, choosing what he liked from these compositions, made a wisdom of his own: much learning, artful knavery [*kakotechne*].

In C H Kahn *The Art and Thought of Heraclitus* 1979 (Cambridge: Cambridge University Press) XXV

61   All things come out of the one, and the one out of all things. Change, that is the only thing in the world which is unchanging.

Aleksandr Ivanovich Herzen 1812-1870

62   The circulation of accurate and meaningful natural science ideas is of vital concern to our age. These are abundant in science but scarce in society. They should be rendered accessible to all...without education in natural science it is impossible to develop a strong intellect.... By placing natural

science at the beginning of a course of education we would cleanse the child's mind of all prejudices; we would raise him on healthful food until the time when, strong intellect...and ready, he discovers the world of man and history which opens the door for direct participation in the issues of the day.

63   Man and science are two concave mirrors continually reflecting each other.
Science and Humanity 1968 (Moscow: Znanie) p 22

Hermann Hesse 1877–1962

64   You treat world history as a mathematician does mathematics, in which nothing but laws and formulae exist, no reality, no good and evil, no time, no yesterday, no tomorrow, nothing but an eternal, shallow, mathematical present.
The Glass Bead Game 1943 (London: Cape) [1972 (Harmonsworth: Penguin) p 158]

65   [The drive to play the glass bead game]...sprang from their deep need to close their eyes and flee from unsolved problems and anxious forebodings of doom into an imaginary world as innocuous as possible.
The Glass Bead Game 1972 (Harmondsworth: Penguin) p 24

David Hilbert 1862–1943

66   One hears a good deal nowadays of the hostility between science and technology. I don't think that is true, gentlemen. I am quite sure that it isn't true, gentlemen. It almost certainly isn't true. It really can't be true. *Sie haben ju gar nichts mit einander zu tun.* [They have nothing whatever to do with one another.]
In J R Oppenheimer *Physics in the Contemporary World* (Cambridge, MA: Harvard University Press)

67   Physics is much too hard for physicists.
In Constance Reid *Hilbert* 1970 (London: Allen & Unwin) p 127

68   *Wir müssen wissen. Wir werden wissen.*
We must know. We will know.
[Speech in Königsberg, 1930. Now on his tomb in Göttingen]
In Constance Reid *Hilbert* 1970 (London: Allen & Unwin) p 196, 220

69   [On the proposed appointment of the first woman professor]
Die Fakultät ist keine Badeanstalt.
The Faculty is not a pool changing room.

70   He who seeks for methods without having a definite problem in mind seeks for the most part in vain.

Joel Henry Hildebrand 1881–1983

71   A child of the new generation
Refused to learn multiplication.
He said. 'Don't conclude
That I'm stupid or rude;
I am simply without motivation.'
Perspectives in Biology and Medicine 1970, Winter, p 230

Harold Hilton

72   It is now generally assumed that matter is not continuous but coarse-grained, i.e. that matter is composed of atoms which are practically indivisible and are situated at very small but not infinitesimal distances apart.
      Mathematical Crystallography 1903 (Oxford: Oxford University Press) p 113

[Sir] Cyril Norman Hinshelwood 1897–1967

73   The creative scientist is in fact usually more concerned with the relations of things to one another than with the precise verbal analysis of what these things are. He seeks a representation of the world which continually grows by an extension or transformation of what is there already. Thus what many scientists are really after is the adventure of discovery itself.
      British Association for the Advancement of Science Presidential Address, Cambridge, 1965

Hippocrates [of Cos] ca 460–ca 357 BC

74   Declare the past, diagnose the present, foretell the future.
      Epidemics Book I, section 11

75   I swear by Apollo the physician, by Asclepius, by Health, by Panacea and by all the gods and goddesses, making them my witnesses, that I will carry out, according to my ability and judgment, this oath and this indenture. To hold my teacher in this art equal to my own parents; to make him partner in my livelihood; when he is in need of money to share mine with him; to consider his family as my own brothers and to teach them this art, if they want to learn it, without fee or indenture; to impart precept, oral instruction, and all other instruction to my own sons, the sons of my teacher, and to indentured pupils who have taken the physician's oath, but to nobody else. I will use treatment to help the sick according to my ability and judgment, but never with a view to injury and wrong-doing. Neither will I administer a poison to anybody when asked to do so, nor will I suggest such a course. Similarly, I will not give a woman a pessary to cause abortion. But I will keep pure and holy both in my life and my art. I will not use the knife, not even, verily, on sufferers from stone but I will give place to such as are craftsmen therein. Into whatsoever houses I enter, I will enter to help the sick, and I will abstain from all intentional wrong-doing and harm, especially from abusing the bodies of man or woman, bond or free. And whatsoever I shall see or hear in the course of my profession, as well as outside my profession in my intercourse with men, if it be what should not be published abroad, I will never divulge holding such things to be holy secret. Now if I carry out this oath, and break it not, may I gain for ever reputation among all men for my life and for my art; but if I transgress it and forswear myself, may the opposite befall me.
      [Great attention is paid to the trade union aspects of the craft and the demarcation between physicians and surgeons]
      The Hippocratic Oath in Harlow Shapley *A Treasury of Science* 1954 (London: Angus & Robertson) p 510

76   Life is short, the Art long, opportunity fleeting, experience treacherous, judgment difficult. The physician must be ready, not only to do his duty himself, but also to secure the co-operation of the patient, of the attendants and of externals.
      Aphorisms I, 1

Hippocrates [School of] ca 460–357 BC

77   It appears to me a most excellent thing, for a physician to cultivate '*pronoia*'. Foreknowing and foretelling in the presence of the sick the past, present, and future (of their symptoms) and explaining all that the patients are neglecting, he would be believed to understand their condition, so that men would have confidence to entrust themselves to his care...Thus he could win just respect and be a good physician. By an earlier forecast in each case he would be more able to tend those aright who have a chance of surviving, and by forseeing and stating who will die, and who will survive, he will escape blame...
Prognostics I

Adolf Hitler 1889–1945

78   Our national policies will not be revoked or modified, even for scientists. If the dismissal of Jewish scientists means the annihilation of contemporary German science, then we shall do without science for a few years.
In E Y Hartshorne *The German Universities and National Socialism* 1937 (London: Allen & Unwin) p 112

Thomas Hobbes 1588–1679

79   Nature (the Art whereby God hath made and governs the World) is by the Art of man, as in many other things, so in this imitated, that it can make an Artificial Animal. For seeing life is but a motion of limbs, the beginning whereof is in some principal part within; why may we not say, that all Automata (Engines that move themselves by springs and wheels as doth a watch) have artificial life For what is the Heart, but a Spring; and the Nerves, but so many Strings; and the joints, but so many Wheels, giving motion to the whole Body, such as was intended by the Artificer? Art goes yet further, imitating that rational and most excellent work of Nature, Man. For by Art is created that great Leviathan called a Commonwealth, or State (in Latin *Civitas*) which is but an artificial man; though of greater stature and strength than the natural, for whose protection and defence it was intended.
Leviathan Introduction, 1651

80   Nature itself cannot err.
Leviathan I, 4, 1651

81   Geometry (which is the only science that it hath pleased God to bestow on mankind)...
Leviathan I, chapter 4, 1651

Eric John Ernest Hobsbawm 1917–

82   There is not much that even the most socially responsible scientists can do as individuals, or even as a group, about the social consequences of their activities.
New York Review of Books XV 19 November 1970

83   War has been the most convenient pseudo-solution for the problems of twentieth-century capitalism. It provides the incentives to modernisation and technological revolution which the market and the pursuit of profit do only fitfully and by accident, it makes the unthinkable (such as votes

for women and the abolition of unemployment) not merely thinkable but practicable, in the field of policy and administration as well as mass murder. What is equally important, it can re-create communities of men and give a temporary sense to their lives by uniting them against foreigners and outsiders. This is an achievment beyond the power of the private enterprise economy, whose characteristic is that it tends to do precisely the opposite, when left to itself.
The Observer Review 26 May 1968

84   All professionals, whether physicists, economists or musicians, live by and for peer judgment, even when they are being paid by people who cannot tell the difference.

August William von Hoffmann 1818–1894

85   I will listen to any hypothesis but on one condition—that you show me a method by which it can be tested.
In R Gregory *The Spirit and Service of Science* 1918 (New York: Macmillan) p 162

Douglas R Hofstadter 1945–

86   Hofstadter's Law: It always takes longer than you expect, even when you take into account Hofstadter's Law.
Goedel, Escher, Bach 1979 (Cambridge, MA: Harvester) p 152

Lancelot Hogben 1895–1875

87   A presbyterian divine once said that a man who plays golf neglects his business, neglects his wife and neglects his God. Many of the elder statesmen of science hold that if a younger one writes a book which can be read painlessly, he neglects his students, neglects his laboratory, and neglects his golf. ...when you become a man you put away childish things. Consequently I do not play golf. In the time which most men of my profession divide between golf and dinner parties, I could write, and have written, several novels, which I have had the good sense to burn.
Preface to *Science for the Citizen* 1938 (London: Allen & Unwin)

88   Science for the Citizen is...also written for the growing number of adolescents, who realize that they will be the first victims of the new destructive powers of science misapplied.
Preface to *Science for the Citizen* 1938 (London: Allen & Unwin)

[Baron] Paul Heinrich Dietrich d'Holbach 1723–1789

89   The unhappiness of man is due to his ignorance of nature.
The System of Nature 1770

90   If a faithful account was rendered of Man's ideas upon Divinity, he would be obliged to acknowledge, that for the most part the word 'gods' has been used to express the concealed, remote, unknown causes of the effects he witnessed; that he applies this term when the spring of the natural, the source of known causes, ceases to be visible: as soon as he loses the thread of these causes, or as soon as his mind can no longer follow the chain, he solves the difficulty, terminates his research, by ascribing it to his gods...When, therefore, he ascribes to his gods the production of some phenomenon...does he, in fact, do anything more than substitute for the

darkness of his own mind, a sound to which he has been accustomed to listen with reverential awe?
Système de la Nature 1770 (London)

91   Theology is but the ignorance of natural causes reduced to a system...[It] is a science that has for its object only things incomprehensible.
Good Sense 1772

Friedrich Hölderlin 1770–1843

92   *Er fliegt, der kühne Geist, wie Adler den Gewittern, weissagend kommenden Göttern voraus.*
Like an eagle before the thunderstorm, the bold spirit flies prophesying before the coming Gods.

Oliver Wendell Holmes 1809–1894

93   A lady's portrait has been known to come out ot the finishing-artist's room ten years younger than when it left the camera. But try to mend a stereo-graph and you will find the difference. Your marks and patches float above the picture and never identify themselves with it. No woman may be declared youthful on the strength of a single photograph; but if the stereo-scopic twins say she is young, let her be so acknowledged.
Sun Painting in *Atlantic Monthly* July 1861

94   Science is the topography of ignorance.
Medical Essays p 211

95   Year after year held the silent toil
That spread his lustrous coil,
Still, as the spiral grew,
He left the past year's dwelling for the new,
Stole with soft step its shining archway through,
Built up its idle door,
Stretched in his last found home, and knew the old no more.
[Not a dwelling but a floatation chamber]
In Peter Douglas Ward *In Search of Nautilus* 1988 (New York: NYAS) p 239

96   The mind, once expanded to the dimensions of larger ideas, never returns to its original size.

Gerald Holton 1922–

97   During a meeting at which a number of great physicists were to give first-hand accounts of their epoch-making discoveries, the chairman opened the proceedings with the remark: 'Today we are privileged to sit side-by-side with the giants on whose shoulders we stand'.
In D J de S Price *Little Science, Big Science* p 1, from Gerald Holton *Am. J. Phys.* 1961 **29** 805

Miroslav Holub 1923–

98   With one bold stroke
He killed the circle, tangent
And point of intersection in infinity.
    On penalty
Of quartering

He banned numbers
From three up.
  Now in Syracuse
He heads a school of philosophers,
Squats on his halberd
For another thousand years
And writes:
One, two
One, two
One, two
One, two.
[Continental armies march 'one, two', rather than 'left, right'.
Holub is a Czechoslovak clinical pathologist whose poems mix scientific, political and
philosophical images]
The Corporal who Killed Archimedes in *New Scientist* 24 July 1969 p 199

Homer ca 900 BC

99   Tell me, O Muse, of the man of many devices...[*polytropon*].
      ['*Polytropic*' is a much better word than 'inter-disciplinary'.]
      First line of the *Odyssey*, introducing Odysseus

Robert Hooke 1635–1703

100   The business and design of the Royal Society is—to improve the knowl-
      edge of natural things, and all useful Arts, Manufactures, Mechanick prac-
      tices, Engynes and Inventions by Experiments—(not meddling with Divin-
      ity, Metaphysics, Moralls, Politicks, Grammar, Rhetorick or Logick)...
      All to advance the glory of God, the honour of the King... the benefit of
      his Kingdom, and the general good of mankind.
      (1663)
      In O R Weld *A History of the Royal Society* London 1848, 1, 146
      Nature 1933 pp 132, 952

101   'The True Theory of Elasticity or Springiness' (1676)—*CEIINOSSITTUU*.
      [Anagram. Revealed in *De Potentia Restitutiva, or of a Spring* (1679) as *UT TENSIO,
      SIC UIS* (as the extension, so the force). One of the ways of establishing priority in a
      discovery]

102   The truth is, the science of Nature has been already too long made only a
      work of the brain and the fancy. It is now high time that it should return
      to the plainness and soundness of observations on material and obvious
      things.
      Micrographia 1665

[Sir] Frederick Gowland Hopkins 1861–1947

103   Life is a dynamic equilibrium in a polyphasic system.
      In J Needham *Order and Life* 1936 (Cambridge: Cambridge University Press) p 132

Rudolf Hoppe 1922–

104   In 1883 Schlegel solved Picasso's problem (how to paint a cow showing all
      features at the same time) by his construction of Schlegel projections.
      Abstract 6.2, *Paul Niggli Symposium* in *ETH Zürich* August 1984

Horace [Q Horatius Flaccus] 65–8 BC

105   *Ac ne forte roges, quo me duce, quolare tuter*

Nullius addictus iurare in verba magistri.
...in the word of no master am I bound to believe.
[Hence *nullius in verba*—the motto of the Royal Society]
[I do not ask, by chance, what leader I follow or what godhead guards me.]

106   *Persicos odi, puer, apparatus.*
Persian luxury, boy, I hate.
[Boy! Those Persians are lousy with apparatus (A L Mackay's translation)]
[See *Perspectives in Biology and Medicine* 1972, Summer, pp 483–90]
Odes I, 38.1

107   *Naturam expellas furca, tamen usque recurret.*
Though you drive away Nature with a pitchfork she always returns.
Epistles I, x, 24

108   *Omne tulit punctum qui miscuit utile dulci.*
He gains everyone's approval who mixes the pleasant with the useful.
Ars Poetica 343

109   The man who makes the experiment deservedly claims the honour and the
reward.
In P C Mangelsdorf *Corn, its Origin, Evolution and Improvement* 1974 (Cambridge,
MA: Harvester Belknap) p ix

Fred Hoyle 1915–

110   I don't see the logic of rejecting data just because they seem incredible.
In D O Edge and M J Mulkay *Astronomy Transformed* 1976 (New York: Wiley) p 432

111   Things are the way they are, because they were the way they were.
Misner, Thorpe and Wheeler attribute this to Thomas Gold

112   [We must] recognise ourselves for what we are—the priests of a not very
popular religion.
Phys. Today April 1968 p 149
Phys. Today 13 February 1987

Hu Shih 1891–1962

113   Contact with strange civilizations brings new standards of value, with which
the native culture is re-examined and re-evaluated, and conscious reforma-
tion and regeneration are the natural outcome.

114   Even my own name bears witness to the great vogue of evolutionism in
China. I remember distinctly the morning when I asked my second brother
to suggest a literary name for me. After only a moment's reflection, he said,
'How about the word *shih* (fitness) in the phrase "survival of the fittest"?'
I agreed, and first using it as a *nom de plume*, finally adopted it in 1910 as
my name.
Living Philosophies, a Series of Intimate Credos 1931 (New York: Simon & Schuster)
p 248

Victor Hugo 1802–1885

115   *Je crois peu à la science des savants bêtes.*
I don't think much of the science of the beastly scientists.

116   Science says the first word on everything, and the last word on nothing.
Things of the Infinite: Intellectual Autobiography transl L O'Rourke, 1907 (New York:
Funk & Wagnalls)

117   There is one thing stronger than all the armies in the world; and that is an idea whose time has come.

David Hume 1711–1776

118   If we take in our hand any volume; of divinity or school metaphysics, for instance; let us ask, 'Does it contain any abstract reasoning concerning quantity or number?' No. 'Does it contain any experimental reasoning concerning matter of fact and existence?' No. Commit it then to the flames: for it can contain nothing but sophistry and illusion.
       Treatise Concerning Human Understanding

119   The supposition that the future resembles the past, is not founded on arguments of any kind, but is derived entirely from habit.
       Treatise of Human Nature I, part iii, section iv

Aldous Leonard Huxley 1894–1963

120   If, O my Lesbia, I should commit,
       Not fornication, dear, but suicide,...
       [Verse on a mistaken belief of the ancients that male corpses floated face up and female corpses face down]
       Second Philosopher's Song in *Collected Poetry of Aldous Huxley* 1971 (London: Chatto & Windus)

121   While I have been fumbling over books
       And thinking about God and the Devil and all,
       Other young men have been battling with the days
       And others have been kissing the beautiful women.
       The Life Theoretic in *Collected Poetry of Aldous Huxley* 1971 (London: Chatto & Windus) p 69

122   The difference between a piece of stone and an atom is that the atom is highly organised, whereas the stone is not. The atom is a pattern, and the molecule is a pattern; but the stone, although it is made up of these patterns, is just a mere confusion. It's only when life appears that you begin to get organisation on a larger scale. Life takes the atoms and molecules and crystals; but, instead of making a mess of them like the stone, it combines them into new and more elaborate patterns of its own.
       Time Must Have a Stop 1945 (London: Chatto & Windus) ch 14

123   Facts are ventriloquist's dummies. Sitting on a wise man's knee they may be made to utter words of wisdom; elsewhere they say nothing, or talk nonsense.
       Time Must Have a Stop 1945 (London: Chatto & Windus) ch 30

124   'If you look up "Intelligence" in the new volumes of the *Encyclopaedia Britannica*', he had said, 'you'll find it classified under the following three heads: Intelligence, Human; Intelligence, Animal; Intelligence, Military. My stepfather's a perfect specimen of Intelligence, Military'.
       Point Counter Point 1928 (London: Chatto & Windus)

125   Science is the reduction of the bewildering diversity of unique events to manageable uniformity within one of a number of symbol systems, and technology is the art of using these symbol systems so as to control and organise unique events. Scientific observation is always a viewing of things through the refracting medium of a symbol system, and technological praxis

is always handling of things in ways that some symbol system has dictated. Education in science and technology is essentially education on the symbol level.
Daedalus Spring 1962 p 279

126   Ends are ape-chosen: only the means are man's.

...

Reason comes running, eager to ratify...
Comes with the calculus to aim your rockets
Accurately at the orphanage across the ocean;
Literature and Science 1963 p 57

127   We have learned that there is an endocrinology of elation and despair, a chemistry of mystical insight, and, in relation to the autonomic nervous system, a meteorology and even, ...an astro-physics of changing moods.
Literature and Science 1963

128   Man cannot live by contemplative receptivity and artistic creation alone. As well as every word proceeding from the mouth of God, he needs science and technology.
Literature and Science 1963 p 35

129   Nineteenth century science discovered the technique of discovery, and our age is, in consequence, the age of inventions...and yet nobody has succeeded in inventing a new pleasure. ...If I were a millionaire, I should endow a band of research workers to look for the ideal intoxicant...the nearest approach to such a drug...is the drug of speed.
Music at Night and Other Essays 1931 (Harmondsworth: Penguin) p 162

130   Our ductless glands secrete among other things our moods, our aspirations, our philosophy of life.
Music at Night and Other Essays 1931 (Harmondsworth: Penguin) p 33

131   'One of those pharmaceutical tragedies', he commented. 'With a course of thiamin chloride and some testosterone I could have made him as happy as a sandboy. Has it ever struck you', he added, 'what a lot of the finest romantic literature is the result of bad doctoring?'
[Huxley explores in his novel, *inter alia*, the consequences of the possibility that man may be a hypertene ape, his development arrested at the foetal stage.]
After Many a Summer 1939 (London: Chatto & Windus) p 234

132   'What is science? Science is angling in the mud—angling for immortality and for anything else that may happen to turn up'.
After Many a Summer 1939 (London: Chatto & Windus) p 232

133   Even if I could be Shakespeare I think that I should still choose to be Faraday.
In W Elsasser *Memoirs of a Physicist in the Atomic Age* (Bristol: Adam Hilger)

Julian Huxley 1887–1975

134   Evolution is the most powerful and the most comprehensive idea that has ever arisen on Earth.
Anthony Smith

Thomas Henry Huxley 1825–1895

135   The great end of life is not Knowledge but Action.
[Marx was saying much the same thing at the time]
Technical Education 1877

136   The great tragedy of Science—the slaying of a beautiful hypothesis by an ugly fact.
[There is an interesting philosophical problem as to the nature of the satisfactions obtained from a scientific insight later found to be fallacious]
Biogenesis and Abiogenesis. Collected Essays viii

137   If all the books in the world except the *Philosophical Transactions* were destroyed, it is safe to say that the foundations of physical science would remain unshaken, and that the vast intellectual progress of the last two centuries would be largely, though incompletely, recorded.

138   If only I could break my leg, what a lot of scientific work I could do.
In Cyril Bibby *T H Huxley* 1959 (London: Watts) p 145

139   It is the customary fate of new truths to begin as heresies and to end as superstitions.
The coming of Age of the Origin of Species in *Science and Culture* xii

140   It looks as if the scientific, like other revolutions, meant to devour its own children; as if the growth of science tended to overwhelm its votaries; as if the man of science of the future were condemned to diminish into a narrow specialist as time goes on.
In *The Essence of T H Huxley* ed Cyril Bibby, 1967 (London: Macmillan) p 234

141   Science is nothing but trained and organized common sense differing from the latter only as a veteran may differ from a raw recruit: and its methods differ from those of common sense only as far as the guardsman's cut and thrust differ from the manner in which a savage wields his club.
The Method of Zadig in *Collected Essays* IV

142   The State lives in a glasshouse, we see what it tries to do, and all its failures, partial or total, are made the most of. But private enterprise is sheltered under good opaque bricks and mortar. The public rarely knows what it tries to do, and only hears of failures when they are gross and patent to all the world.
Administrative Nihilism 1878

143   That fashioning by Nature of a picture of herself, in the mind of man, which we call the progress of Science....
Nature 1869 **1** No. 1, 10

144   This seems to be one of the many cases in which the admitted accuracy of mathematical processes is allowed to throw a wholly inadmissible appearance of authority over the results obtained by them. Mathematics may be compared to a mill of exquisite workmanship, which grinds you stuff of any degree of fineness; but, nevertheless, what you get out depends on what you put in; and as the grandest mill in the world will not extract wheat flour from peascods, so pages of formulae will not get a definite result out of loose data.
[No doubt that this elegant statement of the computer scientist's maxim 'garbage in, garbage out', has still earlier versions]
Quart. J. Geol. Soc., London 1869 **25** 38

145   Try to learn something about everything and everything about something.
Text on his memorial

146   [Asked by Samuel Wilberforce, Bishop of Oxford, whether he traced his descent from an ape on his mother's or his father's side] If the question is

put to me would I rather have a miserable ape for a grandfather or a man highly endowed by nature and possessing great means and influence and yet who employs those faculties and that influence for the mere purpose of introducing ridicule into a grave scientific discussion—I unhesitatingly affirm my preference for the ape.
[Commemorated by a plate in the University Museum, Oxford]
British Association Meeting University Museum, Oxford, 9 September 1860
D Foskett *Nature* 1953 **172** 920

147   [Of the opening ceremony of Johns Hopkins University] It was bad enough to invite Huxley. It were better to have asked God to be present. It would have been absurd to ask them both.
In C Bibby *Scientist Extraordinary—T H Huxley* 1972 (Oxford: Pergamon) p 97

148   [On first reading Darwin's *Origin of Species*] How extremely stupid not to have thought of that!
In K Lorenz *On Agression* 1967 (London: University Paperback) p 237

149   [To the editor of *Science-Gossip* 10 December 1894] Sir, To any one who respects the English language, I think 'Scientist' must be about as pleasing a word as 'Electrocution'. I sincerely trust you will not allow the pages of *Science-Gossip* to be defiled by it. I am, Yours sincerely, Thos. H. Huxley.
In Sydney Ross *Annals of Science* June 1962 **18** 65–85

Ibn Khaldun 1332–1406

1   [Abd-ar-Rahman Abu Zayd ibn Muhammad ibn Muhammad ibn Khaldun]
Geometry enlightens the intellect and sets one's mind right. All its proofs
are very clear and orderly. It is hardly possible for errors to enter into
geometrical reasoning, because it is well arranged and orderly. Thus, the
mind that constantly applies itself to geometry is not likely to fall into
error. In this convenient way, the person who knows geometry acquires
intelligence. It has been assumed that the following statement was written
upon Plato's door: 'No one who is not a geometrician may enter our house'.
The Muqaddimah. An Introduction to History N J Dawood's abridgement of F Rosen-
thal's translation, 1967 (London: Routledge & Kegan Paul) p 378

2   It is a remarkable fact that, with few exceptions, most Muslim scholars
both in the religious and in the intellectual sciences have been non-Arabs.
The Muqaddimah. An Introduction to History N J Dawood's abridgement of F Rosen-
thal's translation, 1967 (London: Routledge & Kegan Paul) p 311

3   Scientific instruction is a craft. This is because skill in a science, knowledge
of its diverse aspects, and mastery of it are the result of a habit.... .The easiest method of acquiring the scientific habit is through acquiring the ability
to express oneself clearly in discussing and disputing scientific problems.
This is what clarifies their import and makes them understandable.
The Muqaddimah. An Introduction to History N J Dawood's abridgement of F Rosen-
thal's translation, 1967 (London: Routledge & Kegan Paul) p 340

4   When the Muslims conquered Persia and came upon an indescribably large
number of books and scientific papers, Sa'd b. Abi Waqqas wrote to 'Umar
b. al-Khattab, asking him for permission to take them and distribute them
as booty among the Muslims. On that occasion 'Umar wrote to him: 'Throw
them into the water. If what they contain is right guidance, God has given
us better guidance. If it is in error God has protected us against it'. Thus,
they [the Muslims] threw them into the water or into the fire, and the
sciences of the Persians were lost and did not reach us.
The Muqaddimah. An Introduction to History N J Dawood's abridgement of F Rosen-
thal's translation, 1967 (London: Routledge & Kegan Paul) p 373

Imanishi Kinji 1902–

5   An individual is merely a component in a system called a species.
A Sibatani 1974

Christopher Kelk Ingold 1893–1970

6   The calculations must have been dreadful...but one structure like this
brings more certainty into organic chemistry than generations of activity
by us professionals.
[On the structures of hexamethyl benzene and hexachlor benzene found by Kathleen
Lonsdale using X-ray crystallographic methods. *Nature* 1928 **1223** 810
Crystallographic Exhibition, The Science Museum, London, 1989

International Business Machines

7   Do not fold, spindle or mutilate.
[A slogan emerging from the student revolution at Berkeley in 1964]
Inscription on IBM card

Isidore of Seville ca 600 AD

8   *Tolle numerum omnibus rebus et omnia pereunt.*
        Take from all things their number and all shall perish.

Jabir ibn Hayyan [Geber] 8th Century

1   The first essential in chemistry is that thou shouldst perform practical work and conduct experiments, for he who performs not practical work nor makes experiments will never attain to the least degree of mastery. But thou, O my son, do thou experiment so that thou mayest acquire knowledge. Scientists delight not in abundance of material; they rejoice only in the excellence of their experimental methods.
Probably The Discovery of Secrets attributed to Geber, 1892 (London: Geber Society)

François Jacob 1920–1977

2   In our universe, matter is arranged in a hierarchy of structures by successive integrations.
The Possible and the Actual 1982 (New York: Pantheon) p 30

3   It is natural selection that gives direction to changes, orients chance, and slowly, progressively produces more complex structures, new organs, and new species. Novelties come from previously unseen association of old material. To create is to recombine.
Science 10 June 1977 **196** 1161–1166

4   In order to drive the individuals towards reproduction, sexuality had therefore to be associated with some other devices. Among these was pleasure...Thus pleasure appears as a mere expedient to push individuals to indulge in sex and therefore to reproduce. A rather successful expedient indeed as judged by the state of the world population.
Evolution and Tinkering in *Science* 10 June 1977 **196** 1161–1166

5   Myths and science fulfil a similar function: they both provide human beings with a representation of the world and of the forces that are supposed to govern it. They both fix the limits of what is considered as possible.
The possible and the Actual 1982 (New York: Random House) p 9

Jalal al-Din [Rumi] 1207–1273

6   Man is God's astrolabe. But it requires an astronomer to know the astrolabe. With that astrolabe what would an ordinary man know of the movements of the circling heavens and the stations of the planet, their influences, transits, and so forth. But in the hands of the astronomer the astrolabe is of great benefit, for he who knows himself knows his Lord. Just as this copper astrolabe is the mirror of the heavens, so the human being is the astrolabe of God. When God causes a man to have knowledge of Him and to know Him and be familiar with Him, through the astrolabe of his own being, he beholds moment by moment and flash by flash the manifestation of God and His infinite beauty and that beauty is never absent from his mirror.
The Discourses of Rumi transl A J Arberry 1961 (London: Murray)

Clive Vivian Leopold James 1939–

7   On my shelves now, collections of aphorisms sit like containers of radioactive material.
The Observer 17 June 1990

William James 1842-1910

8   [Of innovations]...when a thing was new people said 'It is not true'. Later, when its truth became obvious, people said, 'Anyway, it is not important,', and when its importance could not be denied, people said, 'Anyway, it is not new'.
In Lord Ritchie Calder *Leonardo* 1970 (London: Heinemann) p 189

9   Science, like life, feeds on its own decay. New facts burst old rules; then newly developed concepts bind old and new together into a reconciling law.

10   [The new experimental statistical psychology] could hardly have arisen in a country whose natives could be *bored*. Such Germans as Weber, Fechner and Wundt obviously cannot.

Alfred Jarry 1873–1907

11   Oh, but it's like this, look you. I've no grounds to be dissatisfied with my polyhedra; they breed every six weeks, they're worse than rabbits. And its also quite true to say that the regular polyhedra are the most faithful and most devoted to their master, except that this morning the icosahedron was a little fractious, so that I was compelled, look you, to give it a smack on each of its twenty faces. and that's the kind of language they understand. And my thesis, look you, on the habits of polyhedra—it's getting along nicely, thank you, only another twenty-five volumes.
Ubu Cuckolded in *Achras* Act I, Scene I, 1896, but not published until 1944

Thomas Jefferson 1743–1826

12   [On a report by Silliman and Kingsley on a meteorite shower which fell in Weston, CT, in 1807] I could more easily believe that two Yankee professors would lie than that stones would fall from heaven.
1807. *Physics Bulletin* July 1968 **19** 225
In H H Nininger *Our Stone-pelted Planet* 1933 (Boston, MA: Houghton Miflin)

13   If a due participation of office is a matter of right, how are vacancies to be obtained ? Those by death are few: by resignation none.
[Usually quoted as: 'Few die, and none resign']
Letter to a Committee of the Merchants of New Haven, 1801

14   ...freedom, the first-born daughter of science.
Letter to M D Invernois, 6 February 1795

Francis [Lord] Jeffery 1773–1850

15   'Damn the Solar System. Bad light; planets too distant; pestered with comets; feeble contrivance; could make a better myself.'
In H W Tilman *Mischief in Patagonia* (London: Cambridge University Press)

William Stanley Jevons 1835–1882

16   Public opinion, however, is not discriminating and is likely to interpret the agitation for the endowment of science as meaning that science can be had for money.
The Principles of Science Chapter 26

[Pope] John Paul (Woytola) II 1920–

17   Fundamental science is a universal good that all people must be able to cultivate in complete freedom from every form of international servitude or

intellectual colonialism. Basic research must be free with regard to political and economic powers, which must cooperate in its development without impeding its creativity or subjugating it to their own ends. Like any other truth, scientific truth must render account only to itself and to the supreme truth that is God, creator of man and of all things.
Science 14 March 1980 **207** 1165

[Pope] John XXIII 1881–1963

18   Nevertheless, in order to imbue civilisation with sound principles and enliven it with the spirit of the gospel, it is not enough to be illumined with the gift of faith and enkindled with the desire of forwarding a good cause. For this end it is necessary to take an active part in the various organisations and influence them from within. And since our present age is one of outstanding scientific and technical progress and excellence, one will not be able to enter these organisations and work effectively from within unless he is scientifically competent, technically capable and skilled in the practice of his own profession....
Encyclical *Pacem in Terris* 10 April 1963 part 5. Official transl by the Vatican Press Office
[There was a previous Pope John XXIII but he was struck-off!]

Samuel Johnson 1709–1784

19   *Boswell*: 'Is not the Giant's Causeway worth seeing?'
Johnson: 'Worth seeing ? Yes; but not worth going to see.'
Boswell's Life of Johnson 12 October 1779

20   Nay, Madam, when you are declaiming, declaim; and when you are calculating, calculate.
Boswell's Life of Johnson 26 April 1779

21   *Network*: Anything reticulated or decussated, at equal distances, with interstices between the intersections.
Dictionary of the English Language

22   The Sciences having long seen their votaries labouring for the benefit of mankind without reward, put up their petition to Jupiter for a more equitable distribution of riches and honour...A synod of the celestials was therefore convened, in which it was resolved, that Patronage should descend to the assistance of the Sciences.
[Science was then beginning to become a profession]
Rambler 29 January 1751, No 91

23   Sir, I have found you an argument. I am not obliged to find you an understanding.
Boswell's Life of Johnson 26 April 1779

24   [In refutation of the idealism of Bishop Berkeley] 'Johnson struck his foot with mighty force against a large stone, till he rebounded from it, saying, "I refute it thus".'
In Boswell *Life of Johnson* 6 August 1763

25   People have nowadays got a strange opinion that every thing should be taught by lectures. Now, I cannot see that lectures can do as much good as reading the books from which the lectures are taken.
In W Gratzer *Nature* 10 March 1983 **302** 165

Ernest Jones

26 Whenever an individual considers a given (mental) process as being too obvious to permit of any investigation into its origin, and shows resistance to such an investigation, we are right in suspecting that the actual origin is concealed from him—almost certainly on account of its unacceptable nature.
In C H Waddington *The Ethical Animal* 1961 (New York: Athenaeum)

Benjamin Jonson 1572–1637

27 *Surly*: The egg's ordained by Nature to that end, and is a chicken *in potentia*.
Subtle: The same we say of lead and other metals, which would be gold if they had time.
Mammon: And that our art doth further.
The Alchemist II, 2

Maria Jotuni 1880–1943

28 [Finnish writer]
Tiede ei ole ehdotonta eikä tiedemies töineen ulkopuoella tavallisten kuolevaisten virheitä.
Science is not unequivocal, and the scientist with his works is not beyond the mistakes made by ordinary people.

Bertrand de Jouvenel 1903–

29 Year by year we are becoming better equipped to accomplish the things we are striving for. But what are we actually striving for ?
Zukunftspläne. Ritt auf dem Tiger in *Der Spiegel* 1970. No 1–2

Benjamin Jowett 1817–1893

30 One man is as good as another until he has written a book.
In E A Abbott and L Campbell *Life and Letters of B Jowett* 1897, i, p 248

James Joyce 1882–1941

31 I am greatest engineer who ever lived.
In Marshall McLuhan and Quentin Fiore *War and Peace in the Global Village* (New York: Bantam) p 53

32 What in water did Bloom, water lover, drawer of water, water carrier returning to the range, admire?
Its universality: its democratic quality.
Ulysses (Harmondsworth: Penguin) p 592

33 Really it is not I who am writing this crazy book. It is you, and you, and you, and that man over there, and that girl at the next table.
Finnegan's Wake

Thomas Hughes Jukes 1906–

34 Very old are the rocks.
The pattern of life is not in their veins.
When the earth cooled, the great rains
Came and the seas were filled.
Slowly the molecules enmeshed in

Ordered asymmetry.
A billion years passed, aeons of
Trial and error.
The life message took form, a spiral,
A helix, repeating itself endlessly,
Swathed in protein, nurtured by
Enzymes, sheltered in membranes,
Laved by salt water, armored with
Lime.
Shells glisten by the ocean marge,
Surf boils, sea mews cry, and the great wind
Soughs in the cypress.
[Written by a biologist for his own book]
Molecules and Evolution 1966 (New York: Columbia University Press)

Julian ca 331–363

35   [Roman Emperor, Apostate from Christianity to polytheism]
Vicisti O Galilaee!
Thou hast conquered O Galilean!
Attributed last words. In Theodoret, *Hist. Eccl.* iii, 20
[Sometimes applied to Galileo]

Carl Gustav Jung 1875–1961

36   The hypothesis of a collective unconscious belongs to the class of ideas that
people at first find strange but soon come to possess and use as familiar
conceptions. This has been the case with the concept of the unconscious
in general. After the philosophical idea of the unconscious, in the form
presented chiefly by Carus and von Hartman, had gone down under the
overwhelming wave of materialism and empiricism, leaving hardly a ripple
behind, it gradually reappeared in the scientific domain of medical psychology. ...A more or less superficial layer of the unconscious is undoubtedly
personal. I call it the personal unconscious. But this personal unconscious
rests upon a deeper layer, which does not derive from personal experience
and is not a personal acquisition but is inborn. This deeper layer I call the
collective unconscious...it has contents and modes of behaviour that are
more or less the same everywhere and in all individuals. ...The contents of
the collective unconscious, on the other hand, are known as archetypes.
The Archetypes and the Collective Unconscious in *Collected Works* vol 9 (London:
Routledge & Kegan Paul) part 1

37   We can never finally know. I simply believe that some part of the human
Self or Soul is not subject to the laws of space and time.
The Guardian 19 July 1975

Juvenal(is) [Decimus Junius] ca 60–ca 130

38   *Grammaticus, rhetor, geometres, pictor, aliptes, augur, schoenobates, medi-
cus, magus, omnia novit.*
Grammarian, rhetorician, geometer, painter, trainer, soothsayer, rope-
dancer, physician, wizard—he knows everything.
Satires iii, 76

39   [On Alzheimer's disease] Worse by far than any bodily hurt is dementia:
for he who has it no longer knows the names of his slaves or recognises the

friend with whom he has dined the night before, or those whom he had begotten and brought up.

Kalidasa between 200 BC and 400 AD

1   If a professor thinks what matters most
      Is to have gained an academic post
      Where he can earn a livelihood, and then
      Neglect research, let controversy rest,
      He's but a petty tradesman at the best,
      Selling retail the work of other men.
      Malavikagnimitra i.17. In *Poems from the Sanskrit* transl John Brough, 1968 (London:
      Penguin) No 165. ©John Brough, 1968

Mikhail Ivanovich Kalinin 1875–1946

2   Communism is unwittingly being constructed even by the scientist who
      consciously opposes it, but none the less goes on with his work.
      In L R Graham *The Soviet Academy of Sciences and the Communist Party* 1967
      (Princeton, NJ: Princeton University Press) p 44

K'ang Yu-wei 1858–1927

3   In the age of One World, the power of the microscope will be one doesn't
      know how many times greater than that of [the instrument of] today.
      [Viewed through the instrument of today] an ant looks like an elephant.
      [Viewed through the instrument of] the future, the size of a microbe will
      be like that of the great, skyborne *p'eng* bird.
      Ta T'ung Shu: The One-world Philosophy of K'ang Yu-wei transl L G Thompson
      1958 (London: Allen & Unwin)

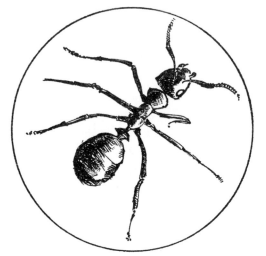

Immanuel Kant 1724–1804

4   Concepts without factual content are empty; sense data without concepts
      are blind.... The understanding cannot see. The senses cannot think. By
      their union only can knowledge be produced.

5   Two things fill the mind with ever new and increasing admiration and awe,
      the oftener and more steadily they are reflected on: the starry heavens
      above me, and the moral law within me.
      Critique of Practical Reason 1788

6   *Zweckmässigkeit ohne Zweck.*
     Purposiveness without a controlling end.
     [A characteristic of biological systems]
     Critique of Judgment 1790

7   Give me matter and I will construct a world out of it.
     Allgemeine Naturgeschichte und Theorie des Himmels, published anonymously in
     1755. 4th edn 1808; republished 1890 (H Ebert)

8   One must understand that the greatest evil that can oppress civilised peo-
     ples derives from wars, not, indeed, so much from actual present or past
     wars, as from the neverending arming for future war. To this end all the
     nation's powers are devoted, as are all those fruits of its culture that could
     be used to build a still greater culture.
     In *New Scientist* 19 May 1990 **126** 67

Peter Leonidovich Kapitsa 1894–1984

9   On the flag of contemporary science there should be written in capital
     letters the word—ORGANISATION.

10   The year that Rutherford died (1938) there disappeared forever the happy
     days of free scientific work which gave us such delight in our youth. Science
     has lost her freedom. Science has become a productive force. She has be-
     come rich but she has become enslaved and part of her is veiled in secrecy.
     I do not know whether Rutherford would continue to joke and laugh as he
     used to.
     Science Policy News, London 1969 **1** No 2, 33

11   [When asked what the significance of the crocodile by Eric Gill on the wall
     of the Royal Society Mond Laboratory was (crocodile, in fact, was Kapitsa's
     name for Rutherford)] The crocodile cannot turn its head. Like science, it
     must always go forward with all-devouring jaws.
     In A S Eve *Rutherford* 1933 (London. Cambridge University Press) p 305

12   Theory is a good thing but a good experiment lasts for ever.
     Experiment, Theory, Practice: Articles and Addresses of P L Kapitsa 1980 (Boston,
     MA: Reidel). *Nature* 11 December 1980 **288** 627

13   [Letter to Stalin, 28 April 1938]
     Comrade Stalin,
     This morning they arrested L D Landau, a co-worker at this Institute...
     [After representations also to Molotov, Kapitsa was allowed the release of
     Landau on his own guarantee]
     26 April 1939
     I beg for the release from custody of the arrested Professor of Physics, Lev
     Davidovich Landau under my personal guarantee.
     I guarantee to the NKVD that Landau will not carry on any counter-
     revolutionary activities whatsoever, against Soviet power in my Institute,
     and I will take upon myself all measures to ensure that also outside the
     Institute he will not carry on any counter-revolutionary works. In the cir-
     cumstance that I notice on the part of Landau any indications in a direction
     likely to harm Soviet power, I will immediately inform the organs of the
     NKVD about them. (signed) P L Kapitsa.
     [Two days later, on the 28 April, that is, about a year after the arrest of

Landau, Kapitsa signed order No 34 of the Institute of Physical Problems: "To restore Comrade Landau, L D, to the list of workers at the Institute of Physical Problems, Academy of Science of USSR with duties as before."]
In P L Kapitsa *Letters about Science* 1989 (Moscow: Moskovskii Rabochii)

14   [Letter to Stalin, 3 October 1945]
Only science and scientists can advance our technology, our economy and the building of our state. ...Therefore, it is high time that comrades of the type of Comrade Beria should begin to learn to respect scientists.
[Beria (1899–1953), was Commissar for Internal Affairs 1938–1945. He was tried and shot in 1953, immediately after the death of Stalin, for gross violations of legality.]
In P L Kapitsa 1989 *Letters about Science* (Moscow: Moskovskii Rabochii) p 400 and p 234

Samuel Karlin 1923–

15   The purpose of models is not to fit the data but to sharpen the questions.
11th R A Fisher Memorial Lecture, Royal Society 20 April 1983

Theodor von Kármán 1881–1963

16   The scientist describes what is: the engineer creates what never was.
Biogr. Mem. FRS 1980 **26** 110

John Keats 1795–1821

17   Do not all charms fly
At the mere touch of cold philosophy?
There was an awful rainbow once in heaven:
We know her woof, her texture; she is given
In the dull catalogue of common things.
Philosophy will clip an angel's wings,
Conquer all mysteries by rule and line,
Empty the haunted air, and gnomed mine
Unweave a rainbow.
Lamia 1820, II, lines 229–37

Keikitsu 1694–1761

18   *Doko no jishaku mo Yoshiwara e muku.*
From wherever it is the magnet points to the Yoshiwara (gay quarter).
Mutamagawa 1754

Thomas à Kempis 1380–1471

19   The humble knowledge of thyself is a surer way to God than the deepest search after science.
De Imitatione Christi part 1, Chapter 3

Maurice George Kendall 1907–1983

20   But I do wish to propound one principle which is, so to speak, a kind of Occam's Electric Razor: We should not invoke any entities or forces to explain mental phenomena if we can achieve an explanation in terms of a possible electronic computer.
Review of the International Statistical Institute 1966 **34** 1

John Fitzgerald Kennedy 1917–1963

21   First, I believe that this nation should commit itself to achieving the goal,
      before this decade is out, of landing a man on the moon and returning
      him safely to earth. No single space project in this period will be more
      exciting, or more impressive to mankind, or more important for the long-
      range exploration of space; and none will be so difficult or expensive to
      accomplish.
      [To Congress, 25 May 1961]
      In John M Logsdon *The Decision to go to the Moon* 1970 (Cambridge, MA: The MIT
      Press)

22   Scientists alone can establish the objectives of their research, but society,
      in extending support to science, must take account of its own needs.
      Address to the National Academy of Sciences, 1963

Hugh Kenner 1923–

23   Each of us carries in his mind a phantom cube, by which to estimate the
      orthodoxy of whatever we encounter in the world of space.
      [This is perhaps more true than Kenner thought. The three semi-circular canals in the
      ear provide us with built-in Cartesian axes]
      Bucky 1973 (New York: Morrow) p 143

24   So writing is largely quotation, quotation newly energised, as a cyclotron
      augments the energies of common particles circulating.
      The Pound Era 1972 (London: Faber) p 126

Johannes Kepler 1571–1630

25   ...Why they are as they are, and not otherwise.
      Mysterium Cosmographicum Preface

26   [Title page of *Tertius Inveniens*] A warning to sundry Theologos, Medicos,
      and Philosophos, in particular to D Philippus Feselius, that they should
      not, in their just repudiation of star-gazing superstition, throw out the
      child with the bath and thus unknowingly act in contradiction to their
      profession.
      In C G Jung and W Pauli *The Interpretation of Nature and the Psyche* 1955 (London:
      Routledge & Kegan Paul) p 180

27   It may well wait a century for a reader, as God has waited six thousand
      years for an observer.
      In David Brewster *Martyrs of Science or the Lives of Galileo, Tycho Brahe, and
      Kepler* 1841

28   *Ubi materia, ibi geometria.*
      Where there is matter, there is geometry.

29   *Certe in Dei Creatoris mente constit Deo coaeterna figurarum harum ver-
      itas.*
      Without doubt the authentic type of these figures exists in the mind of
      God the Creator and shares His eternity.
      The Six-cornered Snowflake (1611). 1966 (Oxford: Oxford University Press) p 36

30   I have only knocked on the door of chemistry and I see how much remains
      to be said.
      The Six-cornered Snowflake (1611). 1966 (Oxford: Oxford University Press)

Charles Franklin Kettering 1876–1958

31   [Vice-president of General Motors]
Bankers regard research as most dangerous and a thing that makes banking hazardous due to the rapid changes it brings about in industry. (Address, 1927.)
In U S National Resources Committee *Technology and Planning* Washington, 1937 pp 5–6
U S Government Report *Technological Trends* 1937 p 63

32   [The future is] where I expect to spend the rest of my life.
In J A Wheeler *Einstein* 1980 (Oxford: Pergamon)

John Maynard Keynes 1883–1946

33   [Of him] No one in our age was cleverer than Keynes nor made less attempt to conceal it.
In R F Harrod *The life of John Maynard Keynes* 1951 (New York: Harcourt Brace Jovanovich)

34   The difficulty lies, not in the new ideas, but in escaping the old ones, which ramify, for those brought up as most of us have been, into every corner of our minds.

35   Newton was not the first of the age of reason. He was the last of the magicians, the last of the Babylonians and Sumerians, the last great mind which looked out on the visible and intellectual world with the same eyes as those who began to build our intellectual inheritance rather less than 10 000 years ago.
[Isaac Newton, a posthumous child born with no father on Christmas Day 1642, was the last wonderchild to whom the Magi could do sincere and appropriate homage.]
Address to the Royal Society Club, 1942

36   The avoidance of taxes is the only intellectual pursuit that still carries any reward.

37   I don't really start until I get my proofs back from the printers. Then I can begin serious writing.
The Guardian 8 June 1983 p 22

Nikita Sergeevich Khrushchev 1894–1971

38   Our party has much experience in working with the intelligentsia. After getting not a few knocks, we gained a correct understanding of many questions. We share this experience with you as friends.
[Speaking in 1958 at the Hungarian Academy of Sciences]
Fundamentals of Marxism-Leninism 1963 (Moscow: Foreign Languages Publishing House), second revised edition

39   Sakharov [in 1958, pleaded] with me not to allow our military to conduct any further tests. He was obviously guided by moral and humanistic considerations. I knew him and was profoundly impressed by him…However, he went too far in thinking that he had the right to decide whether the bomb he had developed could ever be used in the future.
Khrushchev Remembers: The Last Testament 1974 (Boston, MA: Little Brown) p 69

John Tracy Kidder 1945–

40   Each instruction is a precise list of 75 0's and 1's the colourless words of a specific language like early Old English, in which there was no word

for fighting and a poet who wished to convey the idea of a battle had to describe one.
The Soul of a New Machine 1981 (Boston, MA: Little Brown)

Soeren Kiekegaard 1813–1855

41   [For the mysterious quotation earlier attributed to Kierkegaard, see under Musil]

Rudyard Kipling 1865–1936

42   And your rooms at college was beastly—more like a whore's than a man's.
[A dying shipping tycoon explains his will to his more gently nurtured son]
The Mary Gloster in *The Definitive Edition of Rudyard Kipling's Verse* 1940 (London: Hodder & Stoughton)

43   But remember, please the Law by which we live,
We are not built to comprehend a lie,
We can neither love nor pity nor forgive.
If you make a slip in handling us you die.
The Secret of the Machines in *The Definitive Edition of Rudyard Kipling's Verse* 1940 (London: Hodder & Stoughton)

44   For undemocratic reasons and for motives not of State,
They arrive at their conclusions—largely inarticulate.
Being void of self-expression they confide their views to none;
But sometimes in a smoking room, one learns why things were done.
The Puzzler in *The Definitive Edition of Rudyard Kipling's Verse* 1940 (London: Hodder & Stoughton) p 533

45   Nothing in life has been made by man for man's using
But it was shown long since to man in ages
Lost as the name of the maker of it,
Who received oppression and scorn for his wages—
Hate, avoidance, and scorn in his daily dealings—
Until he perished, wholly confounded.
More to be pitied than he are the wise
Souls which foresaw the evil of loosing
Knowledge or Art before time, and aborted
Noble devices and deep-wrought healings,
Lest offence should arise.
Heaven delivers on earth the Hour that cannot be thwarted,
Neither advanced, at the price of a world or a soul, and its Prophet
Comes through the blood of the vanguards who dreamed—too soon—it had sounded.
The Eye of Allah in *Debits and Credits* 1926 (London: Methuen) p 363

Richard Kirwan 1733–1812

46   *Enfin je mets bas les armes, et j'abandonne la phlogistique. Je vois claire-ment qu'il n'y a aucune expérience avérée qui atteste la production de l'air fixe par l'air inflammable pur...Je donnerai moi-même une réfutation de mon 'Essai sur le Phlogistique'.*
At last I put down my arms and I abandon phlogiston. I see clearly that there is no reliable experiment which attests the production of 'fixed air' by

pure 'inflammable air'. I will myself provide a refutation of my own '*Essay on Phlogiston*'.
[A gracious acknowledgment that he was wrong.]
In A F de Fourcroy *Élémens de Chimie* 1791 (Paris: Chez Cuchet). I, preface

Aleksander Isaakovich Kitaigorodskii 1914–

47   A first-rate theory predicts; a second-rate theory forbids and a third-rate theory explains after the event.
Lecture, IUC Amsterdam, August 1975

Spencer Klaw 1920–

48   It is only at long intervals that the researcher enjoys the feeling (or illusion) of solid accomplishment that the administrator can enjoy merely by emptying his in-box.
Science 1969 **163** 60
The New Brahmins' 1968 (New York: Morrow)

Paul Klee 1879–1940

49   *Ich Krystall.*
In D J Haraway *Crystals, Fabrics and Fields* 1976 (London: Yale) p 33

50   And is it not true that even the small step of a glimpse through the microscope reveals to us images which we should deem fantastic and overimaginative if we were to see them somewhere accidentally, and lacked the sense to understand them.
Paul Klee on Modern Art transl Paul Findlay, London, 1948

Aaron Klug 1926–

51   Visiting Americans often think that the tea and coffee breaks are a waste of time but some of them learn better.
The Sunday Times 24 October 1982 p 17

Arthur Koestler 1905–1983

52   Nobody before the Pythagoreans had thought that mathematical relations held the secret of the universe. Twenty-five centuries later, Europe is still blessed and cursed with their heritage. To non-European civilizations, the idea that numbers are the key to both wisdom and power, seems never to have occurred.
The Sleepwalkers 1964 (Harmondsworth: Penguin) p 41

53   To believe entails no desire to know; everybody reads the Bible but who reads Flavius Josephus?
The Yogi and the Commissar 1945 (London: Cape) p 141

54   Among all forms of mentation, verbal thinking is the most articulate, the most complex, and the most vulnerable to infectious diseases. It is liable to absorb whispered suggestions, and to incorporate them as hidden persuaders into the code.
The Act of Creation 1964 (London: Hutchinson) p 177

Roy M Kohn

55   Both horns of a dilemma are usually attached to the same bull: a built-in impediment to understanding psychoses.
[Title of paper]
Perspectives in Biology and Medicine 1970, Summer, p 633

William Lester Kolb 1916–

56   *Science A*: In modern social science usage, the term *Science* denotes the systematic, objective study of empirical phenomena and the resultant bodies of knowledge. It is believed by social scientists that their disciplines are themselves sciences in this sense, and that science as a human activity is itself an object of social science investigation.
A Dictionary of the Social Sciences ed Julius Gould and William L Kolb 1964 (Paris: UNESCO/Tavistock)

Jan Amos Komensky [Comenius] 1592–1670

57   We bid you, then, who are priests in the realm of nature, to press on your labours with all vigour. See to it that mankind is not for ever mocked by a Philosophy empty, superficial, false, uselessly subtle. Your heritage is a fair Sparta; enrich her with fair equipment, and by making a strict examination both of facts and of opinions concerning them, set an example, as you properly may, to politicians and theologians. He was right who said that a contentious philosophy is the parent of a contentious theology: we must therefore say at once and plainly about Politics that the main political theories on which the present rulers of the world support themselves are treacherous quagmires and the real causes of the generally tottering and indeed collapsing condition of the world. It is for you to show that errors are no more to be tolerated, even though they have the authority of long tradition and are drawn from Adam himself; you must show, not only to theologians, but to the politicians themselves, that everything must be called back to Urim and Thummim, I mean to Light and Truth.
The Way of Light Amsterdam, 1668

The Koran

58   We made from water every living thing
Sura 21: 31

50   We have sent down iron, with its mighty strength and diverse uses for mankind, so that Allah may know those who aid Him, though unseen, and help His apostles. Powerful is Allah, and mighty.
Sura *Iron* 57: 25

Korean Proverb

60   *Kwon un sip nyon i yo, sye nun paik nyon i ra.*
Power lasts ten years; influence not more than a hundred.

Damodar Dharmonand Kosambi 1907–1966

61   ...the work remains unique in all Indian literature because of its complete freedom from cant and absence of specious reasoning.
[On the *Arthashastra* of Kautilya]
The Culture and Civilisation of Ancient India 1965 (London: Routledge & Kegan Paul) p 142

Tadeusz Kotarbinski 1886–

62   *Inteligent jest to pasozyt wytwarzacy kulture.*
An intellectual is a parasite that exudes culture.
In *Yugoslav Dictionary of Quotations*

63   *Tam jest potrebny kontroler, gdzie jest potrezebny kontroler kontrolera.*
Where there is the need for a controller, a controller of the controller is also
needed.
[**Juvenal**: *Sed quis custodiet ipsos custodies? Canta est et ab illis incipit uxor.* But
who is to guard the guards themselves? Your wife is as cunning as you and begins with
the guards. (From *Satires* vi, 347.)]

Karl Kraus 1868–1952

64   Science is spectrum analysis. Art is photosynthesis.

David Kresch

65   The understanding of atomic physics is child's play compared with the
understanding of child's play.

Leopold Kronecker 1823–1891

66   *Die ganze Zahl schuf der liebe Gott, alles Übrige ist Menschenwerk.*
God made the integers, man made the rest.
Jahresberichte der deutschen Mathematiker Vereinigung book 2. In F Cajori *A His-
tory of Mathematics* 1919 (London: Macmillan)

[Prince] Peter Alekeseyevich Kropotkin 1842–1921

67   There are not many joys in human life equal to the joy of sudden birth
of a generalization.... . He who has once in his life experienced this joy of
scientific creation will never forget it.

Kuan Yin Tze 8th Century

68   Those who are good at archery learnt from the bow and not from Yi the
Archer. Those who know how to manage boats learnt from boats and not
from Wo (the legendary mighty boatman). Those who can think learnt for
themselves and not from the Sages.
In J Needham *Science and Civilization in China* 1956 (London: Cambridge University
Press) vol II p 73

Kuo Mo-jo [Guo Mo-ro] 1891–1978

69   [Archaeologist and later President of Academia Sinica. A poem on his wel-
come back to Beijing in 1949 after the victory of the PLA]
How much people's blood
Bartered for this pomp?
To think of it tears start,
I laugh—but no sound comes.
[cf Dante *Non vi si pensa quanto sangue costa* in *The Divine Comedy, Paradiso*
Canto XXIX, line 91. Quote by M Taketani]

Otto Vilhelmovich Kuusinen [*et al*] 1881–1964

70   The Marxist–Leninist world outlook stems from science itself and trusts
science, as long as science is not divorced from reality and practice.
Fundamentals of Marxism–Leninism 1963 (Moscow: Foreign Language Publishing
House)

Jean de La Bruyère 1645–1696

1   *Les médécins laissent mourir, les charlatans tuent.*
    The doctors allow one to die; the charlatans kill.
    Les Charactères

Diogenes Laertius 3rd Century BC

2   The first principle of all things is the One. From the One came an indefi-
    nite Two...From the One and the indefinite Two came numbers; and from
    numbers points, lines; from lines, plane figures; from plane figures, solid
    figures, sensible bodies.
    In F M Cornford *Plato and Parmenides* (1957)

Joseph Louis Lagrange 1736–1813

3   [Summarising his life's work] I do not know.

4   The reader will find no figures in this work. The methods which I set forth
    do not require either constructions or geometrical or mechanical reasonings:
    but only algebraic operations, subject to a regular and uniform rule of
    procedure.
    Preface to his *Mécanique Analytique* 1788

5   [On the execution of Lavoisier] Only a moment to cut off that head, and a
    hundred years may not give us another like it.
    In D McKie *Antoine Lavoisier* 1962 (New York: Collier Books) p 262

Imre Lakatos 1922–1974

6   Blind commitment to a theory is not an intellectual virtue: it is an intel-
    lectual crime.
    Philosophical Papers vol 1 1978 (Cambridge: Cambridge University Press) p 1

7   Heuristic is concerned with language-dynamics, while logic is concerned
    with language-statics.

Jean Baptiste Lamarck 1744–1829

8   The most important discoveries of the laws, methods and progress of natu-
    ral thoughts and theories which have passed through the mind of a scientific
    investigator have been crushed in silence and secrecy by his own severe crit-
    icism and adverse examination; that in the most successful instances not a
    tenth of the suggestions, the hopes, the wishes, the preliminary conclusions
    have been realised.

Alphonse Marie Louis Prat de Lamartine 1790–1869

9   It is not I who think, but my ideas which think for me.

Charles Lamb 1775–1834

10 Nothing puzzles me more than time and space; and yet nothing troubles
    me less, as I never think about them.
    Letter to Thomas Manning, 2 January 1806

Frederick William Lanchester 1868–1946

11   [An engineer of genius and a pioneer in the quantization of social phenomena]
The fighting strength of a force may be broadly defined as proportional to the square of its numerical strength multiplied by the fighting value of its individual units.
Aircraft in Warfare 1916 (London: Constable)

12   Versatility is not regarded with favour by the British public.
[1897, his paper on wingtip vortices was rejected by the Physical Society]
1907/1908 *Aerial Flight* [see Royal Society obituary]

Peter Theodore Landsberg 1922–

13   ...the entropy concept that, in spite of its age, 'it has kept an untarnished lustre of novelty, an aura of unplumbed depth. It may well be that it holds further surprises in store for us'.
Entropy and the Unity of Knowledge 1961 (Cardiff: University of Wales Press)

Andrew Lang 1844–1912

14   He uses statistics as a drunken man uses lamp-posts—for support rather than illumination.
Treasury of Humorous Quotations

Rudolph E Langer

15   [Of Fourier] It was, no doubt, partially because of his very disregard for rigor that he was able to take conceptual steps which were inherently impossible to men of more critical genius.
Fourier Series in *Am. Math. Mon.* 1947 supplement to vol **54** pp 1–86. In P J Davis and R Hersh *The Mathematical Experience* 1980 (London: Penguin) p 263

Susanne Katherina Langer 1895–

16   Human life is shot through and through with ritual, as it is also with animalian practices. It is an intricate fabric of reason and rite, of knowledge and religion, prose and poetry, fact and dream...Ritual, like art, is essentially the active termination of a symbolic transformation of experience. It is born in the cortex, not in the 'old brain'; but it is born of an elementary need of that organ, once the organ has grown to human estate.
In C Sagan *The Dragons of Eden* p 61

Ray Lankester 1847–1929

17   The fact that we are able to classify organisms at all in accordance with the structural characteristics which they present, is due to the fact of their being related by descent.
In D'Arcy Thompson *Growth and Form* 1917 (London: Cambridge University Press) p 251

Lao Tze 604–531 BC

18   Clay is moulded to make a vessel, but the utility of the vessel lies in the space where there is nothing...Thus taking advantage of what is, we recognise the utility of what is not.
Tao Te Ching Chapter 41

19   Nature is not human-hearted (anthropomorphic).
Tao Te Ching Chapter 5

20   Of the second-rate rulers, people speak respectfully saying, 'He has done this, he has done that'. Of the first-rate rulers they do not say this. They say: 'We have done it all ourselves'.
[There are dozens of translations of Lao Tze. Some, such as this one, stretch the meaning very far, as the text can mean all things to all men. However, the cumulative effect of the text cannot be mistaken and thus perusal of the whole is recommended. There is indeed some evidence that the dialectical method of Hegel was influenced by Chinese traditions. We do not know who supplied this translation, but the sentiment accurately characterises directors of research projects]
Tao Te Ching Chapter 17

21   A good calculator does not need artificial aids.
Tao Te Ching Chapter 27

Pierre Simon de Laplace 1749–1827

22   *Une intelligence qui, pour un instant donné, connâitrait toutes les forces dont la nature est animée, et la situation respective des êtres qui la composent, si d'ailleurs elle était assez vaste pour soumettre ces données u l'analyse, embrasserait dans la même formula les mouvements des plus grand corps de l'univers et ceux du plus léger atome: rien ne serait incertain pour elle, et l'avenir comme le passé serait présent à ses yeux.*
Given for one instant an intelligence which could comprehend all the forces by which nature is animated and the respective positions of the beings which compose it, if moreover, this intelligence were vast enough to submit these data to analysis, it would embrace in the same formula both the movements of the largest bodies in the universe and those of the lightest atom; to it nothing would be uncertain, and the future as the past would be present to its eyes.
[This conceit is now called 'molecular dynamics']
Oeuvres vol VII. *Théorie Analytique de Probabilité* 1812–1820, Introduction

23   *Napoleon*: 'You have written this huge book on the system of the world
without once mentioning the author of the universe'.
Laplace: 'Sire, I had no need of that hypothesis'.
Later, when told by Napoleon about the incident, Lagrange commented:
'Ah, but that is a fine hypothesis. It explains so many things.'

Eric Larrabee 1922–

24   The only thing wrong with scientists is that they don't understand science. They don't know where their own institutions came from, what forces
shaped them and are still shaping them and they are wedded to an anti-
historical way of thinking which threatens to deter them from ever finding
out.
Science and the Common Reader Commentary, June 1966.
Nature 12 January 1989 **337** 126

Ferdinand Lasalle 1825–1864

25   Whoever obstructs scientific inquiry clamps down the safety valve of public
opinion and puts the state in train for an explosion.
Science and the Workingmen 1863

Emanuel Lasker 1868–1941

26   On the chessboard, lies and hypocracy do not survive long. The creative
combination lays bare the presumption of a lie; the merciless fact, culmi-
nating in a checkmate, contradicts the hypocrite.
[We find that computer programming imposes the same discipline]
Probably *A Manual of Chess* 1932 (London: Constable)

Latin Proverb

27   *Ubi bene ibi patria.*
He makes his home where the living is best.

Johann Caspar Lavater 1741–1801

28   Who in the same given time can produce more than others has *vigour*; who
can produce more and better, has *talents*; who can produce what none else
can, has *genius*.
Aphorisms on Man London, 1788, No 23

David Herbert Lawrence 1885–1930

29   You can't invent a design. You recognise it, in the fourth dimension. That
is, with your blood and your bones, as well as with your eyes.
Phoenix: Art and Morality

30   In Nottingham, that dismal town
Where I went to school and college,
They've built a new university
For a new dispensation of knowledge.
[+ 6 more verses]
Nottingham's New University in *Penguin Book of Unrespectable Verse* ed G Grigson,
1980 (London: Penguin)

31   The Universe is dead for us, and how is it to come alive again? 'Knowledge'
has killed the sun, making it a ball of gas with spots; 'knowledge' has killed

the moon—it is a dead little earth fretted with extinct craters as with smallpox...The world of reason and science...this is the dry and sterile world the abstracted mind inhabits.
In L Wolpert and A Richards (ed) *A Passion for Science* 1988 (Oxford: Oxford University Press) p 1

Edmund Ronald Leach 1910–1989

32 How can a modern anthropologist embark upon a generalisation with any hope of arriving at a satisfactory conclusion?
By thinking of the organisational ideas that are present in any society as a mathematical pattern.
Rethinking Anthropology 1961 (London: Athlone)

Henri Louis Lebesgue 1875–1941

33 In my opinion, a mathematician, in so far as he is a mathematician, need not pre-occupy himself with philosophy—an opinion, moreover, which has been expressed by many philosophers.
In Freeman Dyson *Scientific American* September 1964 **211** 129

Gustave Le Bon 1841–1931

34 At the bidding of a Peter the Hermit millions of men hurled themselves against the East; the words of an hallucinated enthusiast such as Mahomet created a force capable of triumphing over the Graeco–Roman world; an obscure monk like Luther bathed Europe in blood. The voice of a Galileo or a Newton will never have the least echo among the masses. The inventors of genius hasten the march of civilisation. The fanatics and the hallucinated create history.

Henri Le Chatelier 1850–1936

35 Every system in chemical equilibrium, under the influence of a change of any single one of the factors of equilibrium, undergoes a transformation in such direction that, if this transformation took place alone, it would produce a change in the opposite direction of the factor in question. The factors of equilibrium are temperature, pressure and electromotive force, corresponding to the three forms of energy—heat, electricity and mechanical energy.
Recherches sur les Équilibres Chimiques 1888 pp 48, 210. *Comptes Rendus* 1884 **99** 786

Thomas Andrew Lehrer 1928–

36 In one word he told me the secret of success in mathematics: plagiarize...only be sure always to call it please research.
[Song from a very successful gramophone record, ca 1960]
Lobachevski

37 And what is it that put America in the forefront of the nuclear nations? And what is it that made it possible to spend 20 billion dollars of your money to put some clown on the moon? It was good old American know-how, that's what, as provided by good old Americans like Dr Wernher von Braun.
[Gramophone record]
Wernher von Braun in *That Was The Year That Was* (TW3 songs and other songs of the year) 1965

Gottfried Wilhelm Leibnitz 1646–1716

38   The art of discovering the causes of phenomena, or true hypotheses, is like the art of decyphering, in which an ingenious conjecture greatly shortens the road.
New Essays Concerning Human Understanding IV, XII

39   The imaginary number is a fine and wonderful recourse of the divine spirit, almost an amphibian between being and not being.

Stanislaw Lem 1921–

40   In Nature nothing is simple.
Likhtenshtein II p 132

41   Mathematics is...a self-realising infinite field for optional actions, architectural and other labours, for exploration, heroic excursions, daring incursions, surmises.
Non serviam in *A Perfect Vacuum: Perfect Reviews of Non-existent Books* 1971 (New York: Harcourt Brace Jovanovich) p 302

Philip Lenard 1862–1947

42   No entry to Jews and Members of the German Physical Society.
[Notice on his door]
[Nobel Prize, 1905]
In K Mendelssohn *The World of Walther Nernst. The Rise and Fall of German Science* 1973 (London: Macmillan)

Vladimir Ilich Lenin 1870–1924

43   Communism is Soviet power plus the electrification of the whole country.
[Slogan promoting the plan of GOELRO—the State Commission for the Electrification of Russia]
1920

44   Given these economic premises it is quite possible, after the overthrow of the capitalists and bureaucrats, to proceed immediately, overnight to supersede them in the control of production and distribution, in the work of keeping account of labour and products by the armed workers, by the whole of the armed population. (The question of control and accounting must not be confused with the question of the scientifically trained staff of engineers, agronomists and so on. These gentlemen are working today and obey the capitalists; they will work even better tomorrow and obey the armed workers.)
State and Revolution 1917, Chapter 5

45   It is absurd to deny the role of fantasy in even the strictest science.
Polnoe Sobranie Sochinenii 5th edn, vol 29 p 330

46   Socialism is inconceivable without...engineering based on the latest discoveries of modern science.
Left-wing Childishness and the Petty-bourgeois Mentality Moscow, 1968

47   We do not invent, we take ready-made from capitalism; the best factories, experimental stations and academies. We need adopt only the best models furnished by the experience of the most advanced countries.
The Impending Catastrophe 1917

Pope Leo XIII (1878–1903) [Gioacchino Vincenzo Peccei] 1810–1903

48   All the books, which the Church receives as sacred and canonical, are written wholly and entirely, with all their parts, at the dictation of the Holy Ghost; and so far is it from being possible that any error can co-exist with inspiration, that inspiration not only is essentially incompatible with error, but excludes and rejects it as absolutely and necessarily as it is impossible that God Himself, the Supreme Truth, can utter that which is not true.

Leonardo da Vinci 1452–1519

49   The function of muscle is to pull and not to push, except in the case of the genitals and the tongue.
In Edward MacCurdy *The Notebooks of Leonardo da Vinci* vol 1 1938 (London: Cape) Chapter 3

50   *Il sole no si muove.*
The Sun does not move.
Works ed J P and I A Richter (London: Phaidon) 12669a

51   *Nessuna humana investigazione si pio dimandara vera scienzia s'essa non passa per le matematiche dimonstrazione.*
No human investigation can be called real science if it cannot be demonstrated mathematically.
Treatise on Painting transl J P Richter, Chapter 1

52   Shun those studies in which the work that results dies with the worker.
In Edward MacCurdy *The Notebooks of Leonardo da Vinci* vol 1, 1938 (London: Cape) Chapter 1

53   There is no higher or lower knowledge, but one only, flowing out of experimentation.
In Edward MacCurdy *The Notebooks of Leonardo da Vinci* vol 1, 1938 (London: Cape) Chapter 19

54   The science of painting deals with all the colours of the surfaces of bodies and with the shapes of the bodies thus enclosed; with their relative nearness and distance; with the degrees of diminutions required as distances gradually increase; moreover, this science is the mother of perspective, that is, of the science of visual rays.
Paragone London, 1949

55   The supreme misfortune is when theory outstrips performance.

Louis Leprince-Ringuet 1901–

56   *Celui qui trouve ce qu'il cherche fait en général un bon travail d'écolier; pensant à ce qu'il désire, il néglige souvent les signes, parfois minimes, qui aportent autre chose que l'objet de ses prévisions. Le vrai chercheur doit savoir faire attention aux signes qui révéleront l'existence d'un phénomène auquel il ne s'attend pas.*
He who finds what he seeks makes, in general, a good school exercise; intent on what he wants, he often neglects the signs, sometimes minimal, which indicate something else than the object of his attention. The real researcher must pay attention to signs which will reveal the existence of an unexpected phenomenon.
Des Atomes et des Hommes II, Fayard, Paris

R L Lesher and G J Howick

57   Eight hundred life spans can bridge more than 50 000 years. But of these 800 people, 650 spent their lives in caves or worse; only the last 70 had any truly effective means of communicating with one another, only the last 6 ever saw a printed word or had any real means of measuring heat or cold, only the last 4 could measure time with any precision; only the last 2 used an electric motor; and the vast majority or the items that make up our material world were developed within the lifespan of the eight-hundredth person.
Assessing Technology Transfer 1966 *NASA Report* SP-5067 pp 9–10

Leucippus ca 440 BC

58   Not one thing comes to be randomly, but all things from reason and necessity.
In Diels-Kranz *Fragmente der Vorsokratiker* 67B2

Rosario M Levins

59   Confusion evolves into order spontaneously. What God really said was, 'Let there be chaos'.
In H H Pattee (ed) *Hierarchy Theory* 1973 (New York: Braziller) p 127

Claude Levi-Strauss 1908–

60   *La langue est une raison humaine qui a ses raisons, et que l'homme ne connaît pas.*
Language is human reason, which has its internal logic of which man knows nothing.
[Obviously echoes Pascal's dictum: *Le coeur a ses raisons que la raison ne connaît point.*]
La Pensée Sauvage 1962 (Paris: Librairie Plon)

61   No science today can consider the structures with which it has to deal as being more than a haphazard arrangement. That arrangement alone is structured which meets two conditions: that it be a system, ruled by an internal cohesiveness; that this cohesiveness, inaccessible to observation in an isolated system, be revealed in the study of transformations, through which the similar properties in apparently different systems are brought to light.
Leçon Inaugurale in *The Scope of Anthropology* transl S O and R A Paul 1967 (London: Cape) p 27

62   *La pensée mythique n'est pas pre-scientifique; elle anticipe plutôt sur l'état futur d'une science que son mouvement passé et son orientation actuelle montrent progressant toujours dans le même sens.*
Mythical thought is not pre-scientific; rather it anticipates the future state of being a science in that its past movement and its present direction are always in the same sense.
Le Cru et le Cuit Part 4, IIIa (Paris: Plon)

63   *L'ethnologie pourrait se définir comme une technique de dépaysement.*
Ethnology could be defined as a technique of changing the scenery.
Anthropologie Structurale (Paris: Plon) Chapter 6

64   *Le savant d'est pas l'homme qui fournit les vrais réponses; c'est celui qui poses les vraies questions.*

The scientific mind does not so much provide the right answers as ask the right questions.
Le Cru et le Cuit Ouverture, I, (Paris: Plon)

65   All culture can be regarded as an ensemble of symbolic sub-systems, the main ones being language, the conventions of marriage, the economic network, art, science and religion. All these sub-systems aim to express certain aspects of physical reality and of social reality and also the relations which these two types of reality have to each other and the relations which the symbolic sub-systems have to each other. That they can never adequately succeed in this aim is due first to the particular mode of working of each sub-system: they remain always incommensurable; moreover, history tends to make them more different.
Introduction a L'oeuvre de Marcel Mauss. Sociologie et Anthropologie 2nd edn 1960. Introduction p xix

Cecil Day-Lewis 1904–1972

66   And the mind must sweat a poison
...that, discharged not thence
Gangrenes the vital sense
And makes disorder true.
It is certain we shall attain
No life till we stamp on all
Life the tetragonal
Pure symmetry of the brain.
Transitional Poem in *Collected Poems 1929–1933* 1945 (London: Hogarth) p 9

Percy Wyndham Lewis 1884–1957

67   Wherever there is objective truth, there is satire.
Rude Assignment 1950 (London: Hutchinson) p 48

Sinclair Lewis 1885–1951

68   Our American professors like their literature clear, cold, pure and very dead.
Address to the Swedish Academy 1930

William Leybourn 1626–1700

69   But leaving those of the Body, I shall proceed to such Recreations as adorn the Mind; of which those of the Mathematicks are inferior to none.
Pleasure with Profit London, 1694

Georg Christoph Lichtenberg 1742–1799

70   [Professor of Physics in Göttingen]
Where the frontier of science once was is now the centre.

71   A book is like a mirror. If a monkey looks into it no apostle looks out.

72   A book which, above all others in the world, should be forbidden, is a catalogue of forbidden books.

73   All mathematical laws which we find in Nature are always suspect to me, in spite of their beauty. They give me no pleasure. They are merely auxiliaries. At close range it is all not true.
In J P Stern *Lichtenberg* 1959 (London: Thames and Hudson) p 84

74   A good means to discovery is to take away certain parts of a system and
to find out how the rest behaves.
Georg Christoph Lichtenberg's Aphorismen Berlin, 1902–8

[Baron] Justus von Liebig 1803–1873

75   God has ordered all his Creation by Weight and Measure.
[Written over the door of the first chemical laboratory in the world for students.
Giessen, 1842, quoting, as did many others, from the *Wisdom of Solomon* 11:20]

Sam Lilley

76   The form of society has a very great effect on the rate of inventions and a
form of society which in its young days encourages technical progress can,
as a result of the very inventions it engenders, eventually come to retard
further progress until a new social structure replaces it. The converse is
also true. Technical progress affects the structure of society.
Men, Machines and History 1948 (London: Lawrence & Wishart) p 189

Abraham Lincoln 1809–1865

77   Few can be induced to labour exclusively for posterity. Posterity has done
nothing for us.
Speech, 22 February 1842

Charles Augustus Lindberg 1902–1974

78   The tragedy of scientific man is that he has found no way to guide his own
discoveries to a constructive end. He has devised no weapon so terrible that
he has not used it. He has guarded none so carefully that his enemies have
not eventually obtained it and turned it against him. His security today
and tomorrow seems to depend on building weapons which will destroy him
tomorrow.

Frederick Alexander Lindemann [Lord Cherwell] 1886–1957

79   [To Lord De L'Isle (1957)] You know the definition of the perfectly designed
machine...The perfectly designed machine is one in which all its working
parts wear out simultaneously. I am that machine.
In Lord Birkenhead *The Prof in Two Worlds* 1961 (London: Collins) p 332

Eric Linklater [Robert Russell] 1899–1974

80   For the scientific acquisition of knowledge is almost as tedious as the routine
acquisition of wealth.
White Man's Saga 1929 (London: Cape)

Carl Linnaeus 1707–1778

81   *Natura non facit saltus.*
Nature does not make jumps.
Philosophia Botanica 1751, No 77

82   *Lapides crescunt; plantae crescunt et vivunt; animalia crescunt et vivunt et
sentiunt.*
Minerals grow; plants grow and live; animals grow and live and feel.
[In northern countries the phenomenon of frost heave makes the idea, that fields grow
stones, quite sensible.]
Systema Naturae in E A Haeckel 1866 *Generelle Morphologie* Part I p 142

Zbigniew J Lipkowski 1924–

83   To assume, as we all do, that biochemical processes underlie mental activity
      and behaviour does not imply that they are the causal agents but rather
      constitute mediating mechanisms. They are influenced by the information
      inputs we receive from our body and environment and by the subjective
      meaning of that information to us. It is that meaning which largely deter-
      mines what we think, feel and do.
      Can. J. Psychiat. 1989 **34** (3) 249–54

Gabriel Lippmann 1845–1921

84   [On the Gaussian curve, remarked to Poincaré] *Les expérimentateurs
      s'imaginent que c'est un théorème de mathématique, et les mathématiciens
      d'être un fait expérimental.*
      Experimentalists think that it is a mathematical theorem while the math-
      ematicians believe it to be an experimental fact.
      In D'Arcy Thompson *On Growth and Form* 1917 (London: Cambridge University
      Press) I p 121

Walter Lippmann 1889–1974

85   Where all men think alike, no one thinks very much.

Justus Lipsius (Juste Lips) 1547–1606

86   [Quotation carried by numbers of the first volume of the Philosophical
      Magazine, 1798] *Nec aranearum...* The way the spiders weave, you see, is
      none the better because they produce the threads from their own body, nor
      is ours the worse because like bees we cull from the work of others.

Nikolai Ivanovich Lobachevskii 1793–1856

87   [Non-Euclidean geometry] might find application in the intimate sphere of
      molecular attraction.
      ca 1826. *Novie Nachala Geometrii s Polnoi Teoriei Parallelnykh*

John Locke 1632–1704

88   Did we know the mechanical affections of the particles of rhubarb, hemlock,
      opium and a man, as a watchmaker does those of a watch,...we should know
      why rhubarb will purge, hemlock kill and opium make a man sleep.
      1690. *Essay Concerning Human Understanding* in *Nature* 1 September 1983 **305**

Kathleen Lonsdale 1903–1971

89   Any scientist who has ever been in love knows that he may understand
      everything about sex hormones but the actual experience is something quite
      quite different.
      In R Platt *Universities Quarterly* 1963 **17** 327

90   Must relate to peoples needs. Does not mean only engage in applied science,
      does mean always have at back of mind ultimate application is good.
      [Her notes of a conversation in China with Professor Huang, 1955]
      Biogr. Mem. FRS 1976 **21** 447–84

Konrad Lorenz 1903–

91   Historians will have to face the fact that natural selection determined the
      evolution of cultures in the same manner as it did that of species.
      On Aggression 1966 (New York: Harcourt, Brace & World) p 260

92   In nature we find not only that which is expedient, but also everything which is not so inexpedient as to endanger the existence of the species.
On Agression 1966 (New York: Harcourt, Brace & World) p 260
1967 (London: University Paperback) p 132

Alfred J Lotka 1880–1949

93   It would be a strange trick of fate if we, the most advanced product of organic evolution, should be the first of all living species as to forsee its own doom.
[The doom which Lotka forsaw in 1945 was that of human underpopulation!]
Essays on Growth and Form ed Le Gros Clark and P B Medawar 1945 (Oxford: Oxford University Press)

Rudolph Hermann Lotze 1817–1881

94   *Die durchschnittliche Lebensdauer einer physiologischen Wahrheit ist drei bis vier Jahre.*
The average lifespan of a physiological truth is three or four years.

95   Behind the tranquil surface of matter, behind its rigid and regular habits of behaviour, we are forced to seek the glow of a hidden spiritual activity.
Mikrocosmos i 408 (2nd edn)

Ada Augusta [Countess of] Lovelace 1815–1853

96   The distinctive characteristic of the Analytical Engine, and that which has rendered it possible to endow mechanism with such extensive faculties as bid fair to make this engine the executive right-hand of abstract algebra, is the introduction into it of the principle which Jacquard devised for regulating, by means of punched cards, the most complicated patterns in the fabrication of brocaded stuffs. It is in this that the distinction between the two engines lies. Nothing of the sort exists in the Difference Engine. We may say most aptly, that the Analytical Engine *weaves algebraical patterns* just as the Jacquard-loom weaves flowers and leaves. ...In enabling mechanisms to combine together *general* symbols in successions of unlimited variety and extent, a uniting link is established between the operations of matter and the abstract mental processes of the most abstract branch of mathematical science. A new, a vast, and a powerful language is developed for the future use of analysis, in which to wield its truths so that these may become of more speedy and accurate practical application for the purposes of mankind than the means hitherto in our possession have rendered possible...
General Menabrea's Sketch of the Analytical Engine, Invented by Charles Babbage. With Extensive Notes by the Translator October 1842

Amy Lowell 1874–1925

97   Christ! What are patterns for?
Patterns in *The Complete Poetical Works of Amy Lowell* (Boston, MA: Houghton Mifflin)

Lucian ca 115–ca 180

98   *Deus ex machina.*
A God from the machine.
[An allusion to the stage machinery of the theatre, for which see Mary Renault *The Mask of Apollo*]

Titus Carus Lucretius ca 99–ca 55 BC

99   I believe that this world is newly made: its origin is a recent event, not one of remote antiquity. That is why even now some arts are still being perfected: the process of development is still going on. Many improvements have just been introduced in ships. It is no time since organists gave birth to their tuneful harmonies. Yes, and it is not long since the truth about nature was first discovered, and I myself am even now the first who has been found to render this revelation into my native speech.
De Rerum Natura

100   *Nullam rem e gigni divinitus unquam. (Nihil de nihilo fit.)*
Nothing is made from nothing.
[Quoting from *Epicurus, Epist. ad Herodotum*]

101   *Religionum animum nodis exsolvere pergo.*
I am setting out to loose the mind from the knots of religion.
De Rerum Natura I 932

102   Verily, not by design was the world made.

Georg Lukacs 1885–1971

103   Nature is a social category.
History and Class Consciousness transl R Livingstone 1923 (London: The Merlin Press) p 234

Salvador Edward Luria 1912–1991

104   Nature study is the enemy of biology.

105   Jim [J D Watson] makes the very great distinction; if something is not worth doing, it is not worth doing well.
In H Judson *New Yorker* 27 November 1978 p 120

Graham Lusk

106   The work of a man's life is equal to the sum of all the influences he has brought to bear upon the world in which he lives.
Science 1927 **65** 555

Martin Luther 1483–1546

107   *Die Arznei macht kranke, die Mathematik traurige und die Theologie sündhafte Leute.*
Medicine makes people ill, mathematics makes them sad and theology makes them sinful.

André Lwoff 1902–

108   The scientist's art is first of all to find himself a good master.

[Sir] Charles Lyell 1797–1875

109   A scientific hypothesis is elegant and exciting insofar as it contradicts common sense.
[Paraphrased]
Attributed by S J Gould *Ever Since Darwin* 1978 (London: Burnett) p 123

Trofim Denisovich Lysenko 1898–1976

110   The Party, the Government and J V Stalin personally, have taken an un-
flagging interest in the further development of the Michurin teaching.
[This was the session at which Mendelian genetics were condemned and outlawed for
nearly a generation]
Report to the Lenin Academy of Agricultural Sciences Moscow, 31 July–7 August
1948

Thomas Babington [Baron] Macaulay 1800–1859

1   The art which Bacon taught was the art of inventing arts.
Lord Bacon in *Edinburgh Review* July 1837

2   But even Archimedes was not free from the prevailing notion that geometry
was degraded by being employed to produce anything useful. It was with
difficulty that he was induced to stoop from speculation to practice. He was
half ashamed of those inventions which were the wonder of hostile nations,
and always spoke of them slightingly as mere amusements, as trifles in
which a mathematician might be suffered to relax his mind after intense
application to the higher parts of his science.
Lord Bacon in *Edinburgh Review* July 1837

3   Who could deny that a single shelf of a good European library was worth
the whole native literature of India and Arabia?
Minute, 2 February 1835. In G M Trevelyan *Life and Letters of Macaulay* 1881 London
p 290–2

4   [On Francis Bacon]...he who first treated legislation as a science was among
the last English men who used the rack, that he who first summoned
philosophers to the great work of interpreting nature was among the last
Englishmen who sold justice.
Literary Essays in the *Edinburgh Review* July 1837

5   There cannot be a stronger proof of the degree in which the human mind
had been misdirected [by the Schoolmen] than the history of the two greatest events which took place during the middle ages. We speak of the invention of Gunpowder and of the invention of Printing.
Essay on Lord Bacon in the *Edinburgh Review* July 1837

Hugh Macdiarmid [Christopher Murray Grieve] 1892–1978

6   Oor universe is like an e'e
Turned in, man's benmaist hert to see,
And swamped in subjectivity.
   But whether it can use its sicht
To bring what lies withoot to licht
The answer's still ayont my micht.
The Great Wheel in *A Drunk Man Looks at the Thistle* 1926 (Edinburgh: Blackwood)
1926

Ernst Mach 1838–1916

7   The aim of research is the discovery of the equations which subsist between
the elements of phenomena.
Popular Scientific Lectures Chicago 1910 p 205

8   Every statement in physics has to state relations between observable quan-
tities.
[Mach's Principle]

9   The power of mathematics rests on its evasion of all unnecessary thought
and on its wonderful saving of mental operations.
In Freeman Dyson *Scientific American* September 1964 **211** No 3 p 129

10   Science may be regarded as a minimal problem consisting of the completest
presentation of facts with the least possible expenditure of thought.
The Science of Mechanics 9th edn 1942 (LaSalle, IL: Open Court)

11   Archimedes constructing his circle pays with his life for his defective biological adaptation to immediate circumstances.

Antonio Machado 1875–1939

12   *Caminante, no hay camino*
       Se hace camino al andar.
       Traveller, there is no path,
       Paths are made by walking.
       [Theme of popular song in South America]

Alan Lindsay Mackay 1926–

13   How can we have any new ideas or fresh outlooks when 90 per cent of all the scientists who have ever lived have still not died?
       Scientific World 1969 **13** 17–21

14   Like the ski resort full of girls hunting for husbands and husbands hunting for girls, the situation is not as symmetrical as it might seem.
       Lecture, Birkbeck College, University of London, 1964

15   Genotypes never have votes. Phenotypes sometimes do.

16   Magic is a disease of symbols.

Charles Mackay 1814–1889

17   Blessings on Science, and her handmaid Steam
       They make Utopia only half a dream.
       Railways 1846
       The Poetical Works of Charles Mackay 1876 (London: Frederick Warne) p 214

18   Cannon-balls may aid the truth,
       But thought's a weapon stronger;
       We'll win out battles by its aid;—
       Wait a little longer.
       The Good Time Coming
       The Poetical Works of Charles Mackay 1876 (London: Frederick Warne) p 209

19   Truth...and if mine eyes
       Can bear its blaze, and trace its symmetries,
       Measure its distance, and its advent wait,
       I am no prophet—I but calculate.
       The Prospects of the Future
       The Poetical Works of Charles Mackay 1876 (London: Frederick Warne) p 447

[Sir] Halford John Mackinder 1861–1947

20   Knowledge is one. Its division into subjects is a concession to human weakness.
       Proc. R. Geog. Soc. 1887 **9** 141–60

Charles Macklin 1697–1797

21   The law is a sort of hocus-pocus science.

Colin Maclaurin 1698–1746

22   [Writing to James Stirling in 1740]...an unlucky accident has happened to the French mathematicians at Peru. It seems that they were shewing some

French gallantry to the natives' wives, who have murdered their servants destroyed their Instruments and burn't their papers, the Gentlemen escaping narrowly themselves. What an ugly article this will make in a journal.
In Charles Tweedie *James Stirling* 1922 (Oxford: Oxford University Press) pp 90–1

Archibald Macleish 1892–1982

23 Space–time has no beginning and no end.
It has no door where anything can enter.
How break and enter what will only bend?
Reply to Mr Wordsworth

24 It is not for nothing that the scholar invented the Ph.D. thesis as his principal contribution to literary form. The Ph.D. thesis is the perfect image of his world. It is work done for the sake of doing work—perfectly conscientious, perfectly laborious, perfectly irresponsible.
The Irresponsibles

Louis Macneice 1907–1963

25 [Of ancient Athens]
And how can one imagine oneself among them
I do not know.
It was all so unimaginably different.
And all so long ago.

26 A little dapper man but with shiny elbows
And short keen sight, he lived by measuring things
And died like a recurring decimal
Run off the page, refusing to be curtailed;
Died as they say in harness, still believing
In science, reason, progress....
The Kingdom

Pierre Joseph Macquer 1718–1784

27 *La chimie est une science dont l'objet est de reconnaître la nature et les proprietés de tous les corps pars leurs analyses et leur combinaisons.*
Chemistry is a science, the object of which is to recognise the nature and properties of all substances by analysing them and their compounds.
In A F de Fourcroy *Élémens de Chemie* 1791, I, 2

Magna Charta 1215

28 There shall be standard measures of wine, beer, and corn—the London quarter—throughout the whole of our kingdom, and a standard width of dyed, russet and halberject cloth—two ells within the selvedges; and there shall be standard weights also.

Michael Maier 17th Century

29 *Fac ex mare et foemina circulum, inde quadrangulum, hinc triangulum, fac circulum et habebis lapidem philosophorum.*
From a man and a woman make a circle, then a square, then a triangle, finally a circle and you will obtain the Philosophers' Stone.
Scrutinium chymicum Frankfurt 1867
In M Eliade *Forge and Crucible* p 123

Ivan Mikhailovich Maisky 1884–1975

30   [Then Soviet ambassador to the UK] There is no place in the USSR for pure science.
September 1941. *Nature* October 1941 **148** 497

William Harrell Mallock

31   [I have]...the very highest opinion of scandal. It is founded on the most sacred of things—that is, Truth, and it is built up by the most beautiful of things—that is, Imagination.
In O L Dick *Aubrey's Brief Lives* p cxii

André Malraux 1901–1970

32   *Les intellectuels sont comme les femmes; ils s'en prennent aux militaires.*
Intellectuals are like women; they go for the military.
In N Moss *Men who play God* 1970 (London: Penguin) p 241

Thomas Robert Malthus 1766–1834

33   Above all, the Mechanics Institutions, open the fairest prospect that, within a moderate period of time, the fundamentals of political economy will, to a very useful extent, be known to the higher, middle, and a most important portion of the working classes of society in England.
An Essay on the Principles of Population 1798 (London: Everyman)

34   Population, when unchecked, increases in a geometric ratio. Subsistence only increases in an arithmetic ratio.
An Essay on the Principles of Population 1798 (London: Everyman)

35   The passion between the sexes...in every age...is so nearly the same that it may be considered in algebraic language as a given quantity.
An Essay on the Principles of Population 1798 (London: Everyman)

Benois Mandelbrot 1924–

36   Science would be ruined if (like sports) it were to put competition above everything else, and if it were to clarify the rules of competition by withdrawing entirely into narrowly defined specialities. The rare scholars who are nomads-by-choice are essential to the intellectual welfare of the settled disciplines.
Appended to his entry in *Who's Who*

Thomas Mann 1875–1955

37   A great truth is a truth whose opposite is also a great truth.
Essay on Freud 1937

38   I tell them that if they will occupy themselves with the study of mathematics they will find in it the best remedy against the lusts of the flesh.
The Magic Mountain 1927 (London: Secker)

39   Yet each in itself—this was the uncanny, the antiorganic, the life-denying character of them all—each of them was absolutely symmetrical, icily regular in form. They were too regular, as substance adapted to life never was to this degree—the living principle shuddered at this perfect precision, found it deathly, the very marrow of death—Hans Castorp felt he understood now

the reason why the builders of antiquity purposely and secretly introduced minute variations from absolute symmetry in their columnar structures.
[On snowflakes]
The Magic Mountain 1927 (London: Secker)

40   What then was life? It was warmth, the warmth generated by a form-preserving instability, a fever of matter, which accompanied the process of ceaseless decay and repair of protein molecules that were too impossibly ingenious in structure.
The Magic Mountain Chapter 4

Karl Mannheim 1893–1947

41   Even the categories in which experiences are subsumed, collected, and ordered vary according to the social position of the observer.
Ideology and Utopia 1936 (London: Routledge) p 130

Mao Tse-tung 1893–1976

42   ...It is man's social being that determines his thinking. Once the correct ideas characteristic of the advanced class are grasped by the masses, these ideas turn into a material force which changes society and changes the world.
Probably *On Practice*

43   ...hydrogen and oxygen aren't just transformed immediately in any old way into water. Water has its history too.
Mao Tse-tung Unrehearsed ed S Schram 1974 (London: Penguin) p 221

44   The atomic bomb is a paper tiger.... Terrible to look at but not so strong as it seems.
In Anna Louise Strong *A World's Eye View from a Yenan Cave (An Interview with Mao Tse-tung)* 19 April 1947 (New York: Amerasia) pp 122–6

45   Dialectics was interpreted in the past as consisting of three big laws, and Stalin said that it consisted of four big laws. I think there is only one basic law, and that it is the law of contradiction. Quality and quantity, affirmation and negation, phenomenon and essence, content and form, necessity and freedom, possibility and reality, etc, are all unity of opposites.
In Thomas G Hart *The Dynamics of Revolution* 1971 University of Stockholm p 66

46   Knowledge is a matter of science, and no dishonesty or conceit whatsoever is permissible. What is required is definitely the reverse—honesty and modesty.
On Practice included in *The Little Red Book*

47   Letting a hundred flowers blossom and a hundred schools of thought contend is the policy for promoting the progress of the arts and the sciences and a flourishing socialist culture in our land. Different forms and styles in art should develop freely and different schools in science should contend freely. We think that it is harmful to the growth of art and science if administrative measures are used to impose one particular style of art or school of thought and to ban another. Questions of right and wrong in the arts and sciences should be settled through free discussion in artistic and scientific circles and through practical work in these fields. They should not be settled in summary fashion.
On the Correct Handling of Contradictions Among the People included in *The Little Red Book*

48   Marxist philosophy holds that the most important problem does not lie in understanding the laws of the objective world and thus being able to explain it, but in applying the knowledge of these laws actively to change the world.
On Practice included in *The Little Red Book*

49   Natural science is one of man's weapons in his fight for freedom. For the purpose of attaining freedom in society, man must use social science to understand and change society and carry out social revolution. For the purpose of attaining freedom in the world of nature, man must use natural science to understand, conquer and change nature and thus attain freedom from nature.
Speech at the Inaugural Meeting of the Natural Science Research Society for the Border Regions included in *The Little Red Book*

50   Where do correct ideas come from? Do they drop from the skies? No. Are they innate in the mind? No. They come from social practice, and from it alone; they come from three kinds of social practice, the struggle for production, the class struggle and scientific experiment.
Where Do Correct Ideas Come From? included in *The Little Red Book*

51   Tell me, why should symmetry be of importance?
30 May 1974. Tsungdao Lee *Symmetries, Antisymmetries and the World of Particles* 1988 (Washington, DC: Washington University Press)

Ramon Margalef 1919–

52   Probably the hypothesis holds everywhere that the less mature ecosystem feeds the more mature structures around it.
Perspectives in Ecological Theory 1968 (Chicago,IL: University of Chicago Press) p 77

Christopher Marlowe 1564–1593

53   I count religion but a childish toy,
And hold there is no sin but ignorance.
Machiavel in *The Jew of Malta* Prologue

54   Nowe therein of all Sciences (I speak still of humane and according to the humane conceits) is our Poet the Monarch. For he dooth not only show the way but giveth so sweet a prospect into the way as will entice any man to enter into it.
Apologie for Poetrie

Don Marquis 1878–1937

55   An idea isn't responsible for the people who believe in it.
The Sun Dial

Roger Martin du Gard 1881–1958

56   *Le réligion, c'est la science d'autrefois, desséchée, devenue dogme; ce n'est que l'enveloppe d'une explication scientifique depassée depuis longtemps.*
Religion is the science of former times, dried out and turned to dogma; it is only the husk of an outdated scientific explanation.
Jean Barois Le Calme Chapter 2

Groucho Julius Marx 1895–1977

57   What has posterity ever done for me ?
[Attributed]
[Earlier Abraham Lincoln and others]

Karl Marx 1818–1883

58   History itself is an actual part of natural history, of nature's development into man. Natural science will in time include the science of man as the science of man will include natural science: there will be one science.
Writings of the Young Marx on Philosophy and Society ed L D Easton and K H Guddat 1967 (New York: Dobleday) p 312

59   Darwin's book is very important and serves me as a basis in natural science for the class struggle in history. One has to put up with the crude English method of development, of course. Despite all deficiencies not only is the death-blow dealt here for the first time to 'teleology' in the natural sciences, but their rational meaning is empirically explained.

Letter to Lasalle in *Marx-Engels Selected Correspondence, 1846–95* transl Dona Torr 1943 (London: Lawrence & Wishart)

60   The man who draws up a programme for the future is a reactionary.
Letter to Beesley, UK

61   Mankind always takes up only such problems as it can solve; since, looking at the matter more closely, we will always find that the problem itself arises only when the material conditions necessary for its solution already exist or are at least in the process of formation.
[God never sends mouths but sends meat (English Proverb, 1546), but we have few reasons for such confidence in this statement today]
In Karl Marx and Frederick Engels *Selected Works* vol 1 1962 (Moscow: FLPH) pp 361–5

62   Only the working class can...convert science from an instrument of class rule into a popular force.... Only in the Repulic of Labour can science play its proper role.
In Karl Marx and Frederick Engels *On the Paris Commune* 1971 (Moscow: Progress) p 162

63   *Die Philosophen haben die Welt nur verschieden interpretiert, es kommt darauf an, sie zu verändern.* (Eleventh Thesis on Feuerbach.)
The Philosophers have only interpreted the world in various ways; the point, however, is to change it.
[Epitaph on his tomb in Highgate Cemetery, London]
[It seems obvious now that, in order to change the world in the direction you wish, you first have to understand it]

64   The product of mental labour—science—always stands far below its value, because the labour-time necessary to reproduce it has no relation at all to the labour-time required for its original production.
Theories of Surplus Value

65   We know only a single science, the science of history. History can be con-
templated from two sides, it can be divided into the history of nature and
the history of mankind. However, the two sides are not to be divided off; as
long as men exist the history of nature and the history of men are mutually
conditioned.
The German Ideology

66   The bourgeoisie has stripped of its halo every occupation hitherto honoured
and looked up to with reverent awe. It has converted the physician, the
lawyer, the priest, the poet, the man of science, into its paid wage labourers.
The Communist Manifesto

67   At the entrance to science, as at the entrance to Hell, there should be
posted the demand:'Here the spirit should be firm. Here the promptings of
fear should not be heeded'.
R V Bogdanov

68   Lucretius is the truly Roman heroic poet; his heroes are the atoms, inde-
structible, impenetrable, well-armed, lacking all qualities but these; a war
of all against all, the stubborn form of eternal substance. Nature without
gods, gods without a world.
Ergänzungsband in *Marx–Engels Gesamtausgabe* vol 1 1968 (Berlin: Dietz) p 170

69   One basis for life and another for science is *a priori* a lie.
EPM, 1884. *Marx–Engels Gesamtausgabe* vol 1 1968 (Berlin: Dietz) p 121

70   The dealer in minerals sees only their money value, not the beauty or the
special character of the minerals; he has no mineralogical sense.
In L H Gould *Marxist Glossary* Sydney, 1947

71   Reason has always existed, but not always in a reasonable form.

John Masefield 1878–1967

72   ...Harwell Man's perpetual treasure-trove....
The Collected Poems of John Masefield 1923 (London: Heinemann)

Matsushita Electrical Company

73   For the building of a new Japan
Let's put mind and strength together,
Doing our best to promote production,
Sending our goods to the peoples of the world,
Endlessly and continuously,
Like water gushing from a fountain.
Grow industry, grow, grow, grow,
Harmony and sincerity. Matsushita Electrical.
[Company hymn. Sung like a school song, on official occasions, to promote *esprit de corps*]
In F L K Hsu *Iemoto, The Heart of Japan* 1975 (New York: Wiley) p 211

Bernd T Matthias

74   [Professor of Physics, University of California at San Diego.]
If you see a formula in the *Physical Review* that extends over a quarter of
a page, forget it. It's wrong. Nature isn't that complicated.

William Somerset Maugham 1874–1965

75   It is bad enough to know the past; it would be intolerable to know the future.
In Richard Hughes *Foreign Devil* 1972 (London: Deutsch) p 209

André Maurois 1885–1967

76   *Il y a une évidente parenté entre la maxime du moraliste et la loi scientifique. La loi est une pensée verifée par des experiences méthodiques; la maxime est une loi suggerée par l'experience vécue. L'une et l'autre tentent d'exprimer un rapport entre deux éléments.*
There is an evident kinship between the maxim of the moralist and a scientific law. The law is a thought verified by systematic experiments; the maxim is a law suggested by experience. Both try to express a relationship between two elements.
Pensée Scientifique et Oeuvre Littéraire de Jean Rostand 1968 (Paris: Larousse) p 25

Osborne Henry Mavor [James Bridie] 1888–1957

77   Youth is greatly interested in psychology, or the study of the Soul. The Soul first began to attract attention about a quarter of a century ago when Dr Freud at Heaven's command arose from out the azure main. He discovered a thing called the Unconscious, which is a kind of consciousness; just as non-existence is a kind of existence.... Psychology is singularly adapted for the use of youth, because it has a large vocabulary and is, in its essence, a grand, riotous, complicated, smutty story. It also moves along tram lines and is dressed in uniform. Youth, terrified by the roaring, multitudinous manifestations of life all around it, feels safe with psychology.
Mr Bridie's Alphabet for Little Glasgow Highbrows

James Clerk Maxwell 1813–1879

78   ...that, in a few years, all great physical constants will have been approximately estimated, and that the only occupation which will be left to men of science will be to carry these measurements to another place of decimals.
[Maxwell himself categorically rejected this view and was attacking it]
Scientific Papers October 1871 **2** 244

79   In the very beginning of science,
The parsons, who managed things then,
Being handy with hammer and chisel,
Made gods in the likeness of men;
Till Commerce arose and at length
Some men of exceptional power
Supplanted both demons and gods by
The atoms, which last to this hour.
[+ further verses]
Notes of the President's Address, 1874
British Association for the Advancement of Science

80   The only laws of matter are those which our minds must fabricate, and the only laws of mind are fabricated for it by matter.
In J G Crowther *British Scientists of the Nineteenth Century* 1935 (Harmondsworth: Penguin) p 316

81   I come from fields of fractured ice,
      Whose wounds are cured by squeezing,
      Melting they cool, but in a trice,
      Get warm again by freezing.
      To the Chief Musician upon Nabla

82   The use of diagrams is a particular instance of that method of symbols
      which is so powerful an aid in the advancement of science.
      In D'A W Thompson *Growth and Form* 1948 (Cambridge: Cambridge University Press)
      p 995

83   Thus science strips off, one after the other, the more or less gross materi-
      alisations by which we endeavour to form an objective image of the soul,
      till men of science, speculating, in their non-scientific intervals, like other
      men, on what science may possibly lead to, have prophesied that we should
      soon have to confess that the soul is nothing else than a function of certain
      complex material systems.
      Review of B Stewart and P G Tait *Paradoxical Philosophy* in *Nature*. Also Maxwell's
      Collected Works 1869, II, p 760

84   By referring everything to the purely geometrical idea of the motion of an
      imaginary fluid, I hope to attain generality and precision, and to avoid the
      dangers arising from a premature theory professing to explain the causes
      of phenomena. If the results of mere speculation which I have collected
      are found to be of any use to experimental philosophers, in arranging and
      interpreting their results, they will have served their purpose, and a mature
      theory, in which physical facts will be physically explained, will be formed
      by those who, by interrogating Nature herself, can obtain the only true
      solution of the questions which the mathematical theory suggests.
      On Faraday's Laws of Force 1856

85   The simplest case is that of five points in space with their ten connecting
      lines, forming ten triangular faces enclosing five tetrahedrons. By joining
      the five points which are the centres of the spheres circumscribing these five
      tetrahedrons, we have a reciprocal figure of the kind described by Professor
      Rankin in the *Phil. Mag.* of February 1864; and forces proportional to the
      areas of the triangles of one figure, if applied along the corresponding lines
      of connexion of the other figure, will keep its points in equilibrium.
      On reciprocal polyhedra in *Collected Works* 1869

86   The University of Cambridge, in accordance with that law of its evolution,
      by which, while maintaining the strictest continuity between the successive
      phases of its history, it adapts itself with more or less promptness to the
      requirements of the times, has lately instituted a course of Experimental
      Physics.
      [His inaugural lecture as the first Professor of Experimental Physics]
      Collected Works 1869, p 241

Robert McCredie May 1936–

87   [On being reproached for an impossible attribution of an anecdote about
      J B S Haldane] Mundane constraints of time and space do not apply to
      stories about Oxford.
      Nature 26 October 1989 **341** 695

88   Not only in research, but in the everyday world of politics and economics, we would all be better off if more people realised that simple systems do not necessarily possess simple dynamical properties.
      Nature 1976 **261** 459–67

Vladimir Vladimirovich Mayakovsky 1893–1930

89   Chicago: City
      Built upon a screw!
      Electro-dynamical-mechanical city!
      Spiral shaped—
      On a steel disc—
      At every stroke of the hour
      Turning itself round!

Joseph McCabe 1867–1955

90   Millions upon millions of [ganglionic] cells are woven into the gray bed or cortex of the brain, with which intelligence is associated. To say that all this complexity is only for the purpose of letting in a spiritual principle from another world is gratuitous in the extreme.
      The Origin of Life: A Reply to Sir Oliver Lodge 1906

Warren S McCulloch 1898–1969

91   Don't bite my finger—look where it's pointing.
      [The pioneer of modelling the neurons and their operation]
      In Stafford Beer *Platform for Change* 1975 (Chichester: Wiley)

92   To the theoretical question, Can you design a machine to do whatever a brain can do? the answer is this: If you will specify in a finite and unambiguous way what you think a brain does do with information, then we can design a machine to do it. Pitts and I have proved this constructively. But can you say what you think brains do?

McGregor [pseudonym for Jacques Monod 1910–1976]

93   Each of Science's conquests is a victory of the absurd.
      Inaugural Lecture, 3 November 1967 *From Biology to Ethics* Salk Institute Occas. Paper 1, 1969 p 19

John McLeod and John Osborn

94   ...in real life mistakes are likely to be irrevocable. Computer simulation, however, makes it economically practical to make mistakes on purpose. If you are astute, therefore, you can learn much more than they cost. Furthermore, if you are at all discreet, no one but you need ever know you made a mistake.
      Natural Automata and Useful Simulations ed H H Pattee *et al* 1966 (London: Macmillan) p 128

Herbert Marshall McLuhan 1911–

95   When this circuit learns your job, what are you going to do?
      In Allen Lane *The Medium is the Massage* 1967 (Harmondsworth: Penguin) p 20

[Sir] Peter Brian Medawar 1915–1987

96   Considered in its entirety, psychoanalysis won't do. It is an end product, moreover, like a dinosaur or a zeppelin; no better theory can ever be erected

on its ruins, which will remain for ever one of the saddest and strangest of all landmarks in the history of twentieth century thought.
The Hope of Progress 1972 (London: Methuen) p 68

97   No scientist is admired for failing in the attempt to solve problems that lie beyond his competence. The most he can hope for is the kindly contempt earned by the Utopian politician. If politics is the art of the possible, re-search is surely the art of the soluble. Both are immensely practical-minded affairs. Good scientists study the most important problems they think they can solve. It is, after all, their professional business to solve problems, not merely to grapple with them.
The Art of the Soluble 1967 (London: Methuen) p 7

98   [On Teilhard de Chardin *The Phenomenon of Man*] The greater part of it, I shall show, is nonsense, tricked out with a variety of tedious metaphysical conceits, and its author can be excused of dishonesty only on the grounds that before deceiving others he has taken great pains to deceive himself.
The Art of the Soluble 1967 (London: Methuen) p 71

99   Among scientists are collectors, classifiers, and compulsive tidiers-up; many are detectives by temperament and many are explorers; some are artists and others artisans. There are poet–scientists and philosopher–scientists and even a few mystics.
The Art of the Soluble 1967 (London: Methuen)

100   People who write obscurely are either unskilled in writing or up to mischief.
Science and Literature in *Pluto's Republic* 1984 (Oxford: Oxford University Press) p 52

101   The case I shall find evidence for is that when literature arrives, it expels science.
Science and Literature in *Pluto's Republic* 1984 (Oxford: Oxford University Press)

102   To deride the hope of progress is the ultimate fatuity, the last word in poverty of spirit and meanness of mind.
The Hope of Progress 1973 (New York) p 137

[Sir] **Peter Brian Medawar** 1915–1987 and **Lady Jean Shinglewood Medawar**

103   Medical scientists use the word 'iatrogenic' to refer to disabilities that are the consequence of medical treatment. We believe that some such word might be coined to refer to philosophical difficulties for which philosophers themselves are responsible.
Aristotle to Zoos 1983 (Cambridge, MA: Harvard University Press) p 244

Charles Edward Kenneth Mees 1882–1960

104   The best person to decide what research shall be done is the man who is doing the research. The next best is the head of the department. After that you leave the field of best persons and meet increasingly worse groups. The first of these is the research director, who is probably wrong more than half the time. Then comes a committee which is wrong most of the time. Finally there is a committee of company vice-presidents, which is wrong all the time.
[Uttered in 1935. Mees was Research Director of Kodak]
Biographical Memoirs of Fellows of the Royal Society 1961 **7** 182

[The] Meiji Emperor of Japan 1868–1911

105   Knowledge shall be sought from all parts of the world in order to strengthen the Imperial system.
The Charter Oath 6 April 1868

Herman Melville 1819–1891

106   Captain Ahab: 'My means are sane, my motive and my object mad'.
Moby Dick 1851

107   Chemist, you breed
In orient climes each sorcerous weed
That energises dream—
The New Zealot to the Sun

Mencius 372–289 BC

108   King Hu of Leang said: 'I wish quietly to receive instruction'. Mencius replied: 'Is there any difference between killing a man with a stick and killing a man with a sword?' The king said: 'There is no difference'. 'Is there any difference between killing a man with a sword and killing a man with a system of government?'. He replied: 'There is no difference'.
Mencius I.4, 1–3

Henry Louis Mencken 1880–1956

109   Science, at bottom, is really anti-intellectual. It always distrusts pure reason and demands the production of the objective fact.
Notebooks Minority Report I

Dmitri Ivanovich Mendeleev 1834–1907

110   I am not afraid of the admission of foreign, even of socialistic ideas into Russia, because I have faith in the Russian people who have already got rid of the Tatar domination and the feudal system.

111   There will come a time, when the world will be filled with one science, one truth, one industry, one brotherhood, one friendship with nature...this is my belief, it progresses, it grows stronger, this is worth living for, this is worth waiting for.
In Yu A Urmantsev *The Symmetry of Nature and the Nature of Symmetry* 1974 (Moscow: Mysl) p 49

Menachem Mendl [Rabbi, of Kursk]

112   If I am I because I am I, and you are you because you are you, then I am I and you are you. However, if I am I because you are you, and you are you because I am I, then I am not I and you are not you.
[Separability of the wave-function?]

Norman David Mermin 1935–

113   [Co-author of 'the world's funniest solid-state physics text'—*ipse dixit*]
The EPR experiment is as close to magic as any physical phenomenon I know of, and magic should be enjoyed. Whether there is physics to be learned by pondering it is less clear.
A Einstein, B Podolsky and N Rosen *Physics Review* 1935 **47** 777
In *Physics Today* April 1985 pp 2–11

114   Bridges would not be safer if only people who knew the proper definition of a real number were allowed to design them.
Topological Theory of Defects in *Review of Modern Physics* July 1979 **51** No 3 591–648

Robert King Merton 1910–

115   Most institutions demand unqualified faith; but the institution of science makes skepticism a virtue.
Social Theory and Social Structure 1962 (New York: Free Press) p 547

116   Science is public, not private, knowledge.
Science, Technology and Society in Seventeenth-century England 1938 (New York: Fertig) p 219

[Prince] Clemens Wenzel Lothar Metternich-Winneburg 1773–1859

117   [To the Austrian Ambassador in London for transmission to King George IV, 1825] There is one matter which I beg you to bring to the King's notice yet again before your departure; this is the proposed foundation of a university of London. You have my authority to tell His Majesty of my absolute conviction that the implementation of this plan would bring about England's ruin.
In G de Bertier De Sauvingny *Metternich and his Times* in *University of London Bulletin* No 2, January 1972

Albert Michaelson 1852–1956

118   Physical discoveries in the future are a matter of the sixth decimal place.

Ivan Vladimirovich Michurin 1855–1935

119   We must not wait for favours from Nature; our task is to wrest them from her.
[Slogan of the Lysenkoist school]
Short Dictionary of Philosophy Moscow 1955

Veljko Mićunović

120   Military strength is unfortunately still the first beneficiary and main indicator of the progress of science and technology in the Soviet Union.
Moscow Diary 1980 (London: Chatto & Windus)

Ludwig Mies van der Rohe 1886–1969

121   Less is more.
[The architect affirms the positive virtue of shaving with Occam's Razor (q.v.)]
Obituary in *The Times* 19 August 1969

122   The long path from material through function to creative work has only one goal: to create order out of the desperate confusion of our time.

John Stuart Mill 1806–1873

123   The habit of analysis has a tendency to wear away the feelings.
Autobiography v

Henry Miller 1891–1977

124   The wallpaper with which the men of science have covered the world of reality is falling to tatters.
The Tropic of Cancer 1934 (London: Calder)

John Milton 1608–1674

125   *The Argument*: Adam inquires concerning celestial motions; is doubtfully
      answered, and exhorted to search rather things more worthy of knowledge.
      [Kepler died in 1630]
      Heading in *Paradise Lost* 1667 Book VIII

126   Behold now this vast City: a city of refuge, the mansion house of liberty,
      encompassed and surrounded with His protection; the shop of war hath
      not there more anvils and hammers waking, to fashion out plates and in-
      struments of armed justice in defence of beleaguered Truth, than there be
      pens and hands there, sitting by their studious lamps, musing, searching
      revolving new motions....
      [Describing London during the Civil War]
      Areopagitica

127   Eccentric, intervolved, yet regular
      Then most, when most irregular they seem;
      And in their motions harmony divine.
      Paradise Lost 1667, Book V, 623

128   From Man or Angel the great Architect
      Did wisely to conceal, and not divulge,
      His secrets, to be scanned by them who ought
      Rather admire. Or, if they list to try
      Conjecture, he his fabric of the Heavens
      Hath left to their disputes—perhaps to move
      His laughter at their quaint opinions wide
      Hereafter, when they come to model Heaven
      And calculate the stars: how they will wield
      The mighty frame: how build, unbuild, contrive
      To save appearances; how gird the Sphere
      With Centric and Eccentric scribbled o'er,
      Cycle and Epicycle, Orb in Orb.
      Paradise Lost 1667, Book VIII, 72

129   [Mulciber, the architect of the great palace of Pandemonium, had been
      thrown out of Heaven]
      ...nor aught availed him now
      To have built in Heaven high towers; nor did he 'scape
      By all his engines, but was headlong sent
      With his industrious crew to build in Hell.
      [The theory of the takeover which ousted Satan and his colleagues is entertainingly
      discussed by Anthony Jay in *Management and Machiavelli* 1970 (London: Penguin)]
      Paradise Lost 1667, Book I, 748

130   [The Tree of Knowledge] O Sacred, Wise and Wisdom-giving Plant,
      Mother of Science.
      Paradise Lost 1667, Book IX, 679

131   For hot, cold, moist, and dry, four champions fierce,
      Strive here for mastery, and to battle bring
      Their embryon atoms;....
      Paradise Lost 1667, Book II, 898–900

132   O Sacred, Wise and Wisdom-giving Plant,
       Mother of Science, Now I feel thy Power,
       Within me cleere, not onely to discerne
       Things in their Causes, but to trace the wayes
       Of highest Agents, deemed hoever wise.
       Paradise Lost 1667 Book IX, 679–83

133   For such kind of borrowing as this, if it be not bettered by the borrower,
       among good authors is accounted Plagiaré.
       Iconoclastes 24

134   So easy it seemed
       Once found, which yet unfound most would have thought impossible.

135   [Who, in 1638, visited the blind Galileo in Arcetri] There [in Catholic Italy]
       it was that I found and visited the famous Galileo grown old, a prisoner to
       the Inquisition, for thinking in Astronomy otherwise than the Franciscan
       and Dominican Licensers thought.
       Areopagitica. For the Liberty of Unlicenc'd Printing 1644

Herman Minkowski 1864–1909

136   The views of space and time which I wish to lay before you have sprung
       from the soil of experimental physics, and therein lies their strength. They
       are radical. Henceforth space by itself, and time by itself, are doomed to
       fade away into mere shadows, and only a kind of union of the two will
       preserve an independent reality.
       September 1908. In L D Henderson *The Fourth Dimension and Non-Euclidean Ge-
       ometry in Modern Art* 1983 (Princeton, NJ: Princeton University Press) p 356

Marvin Lee Minsky 1927–

137   Logic doesn't apply to the real world.
       The Mind's I ed D R Hofstadter and D C Dennett 1981(Cambridge, MA: Harvester)
       p 343

Charles W Misner 1932–, Kip S Thorne 1940– and John Archibald Wheeler 1911–

138   Physics is simple only when analysed locally.
       Gravitation 1972 (San Francisco, CA: Freeman) p 4

139   Space acts on matter, telling it how to move. In turn, matter reacts back
       on space telling it how to curve.
       or: Geometry tells matter how to move, and matter tells geometry how to
       curve.
       Gravitation 1972 (San Francisco, CA: Freeman) pp 5 and 130

George John Mitchell 1933–

140   [Senator, Dem., Maine] Although he is regularly asked to do so, God does
       not take sides in American politics.
       Senate Hearing 18 July 1985. *Los Angeles Times* 14 July 1987 p 13

Alwyn Mittasch 1869–1953

141   Chemistry without catalysis, would be a sword without a handle, a light
       without brilliance, a bell without sound.
       Journal of Chemical Education 1948 531–2

[Sir] Walter Hamilton Moberley 1881–1974

142   For God's sake, stop researching for a while and begin to think.
The Crisis in the University 1949 (London: Student Christian Movement Press) p 183

Jacob Moleschott 1822–1893

143   [Life is] woven out of air by light.
Perspectives in Biology and Medicine 1972, Winter, p 208

Jacques Monod 1910–1977

144   *Il y a des systèmes vivants; il n'ya pas de 'matière' vivante.*
There are living systems; there is no 'living matter'.
Inaugural Lecture, College de France, Chaire de Biologie Moleculaire, 3 November 1967

145   Language may have created man, rather than man language.
Inaugural Lecture, College de France, Chaire de Biologie Moleculaire, 3 November 1967

146   The ancient covenant is in pieces; man knows at last that he is alone in the universe's unfeeling immensity, out of which he emerged only by chance. His destiny is nowhere spelled out, nor is his duty. The kingdom above or the darkness below: it is time for him to choose.
[Closing words] *Chance and Necessity* 1970 (Paris: Editions du Seuil)

147   In science, self-satisfaction is death. Personal self-satisfaction is the death of the scientist. Collective self-satisfaction is the death of the research. It is restlessness, anxiety, dissatisfaction, agony of mind that nourish science.
New Scientist 17 June 1976 **70** 680

Michel Eyquem Montaigne 1533–1592

148   If, by being overstudious, we impair our health and spoil our good humour…let us give it over.
Essais

149   Science without conscience is but death of the soul.
Essais

150   Whenever a new discovery is reported to the scientific world, they say first, 'It is probably not true.' Thereafter, when the truth of the new proposition has been demonstrated beyond question, they say, 'Yes, it may be true, but it is not important.' Finally, when sufficient time has elapsed fully to evidence its importance, they say, 'Yes, surely it is important, but it is no longer new.'

Charles-Louis de Secondat Montesquieu 1689–1755

151   [To Sophie of Hannover about G W Leibnitz] It is rare to find learned men who are clean, do not stink and have a sense of humour.
Liselotte in *Les Lettres Persanes* LXVI, 30 July 1705
But we also find: The Duchess of Orléans [commenting favourably on Leibnitz]
It's so rare for intellectuals to be smartly dressed, and not to smell, and to understand jokes.
In G MacDonald Ross *Leibnitz* 1984 (Oxford: Oxford University Press)

[Sir] Thomas More 1478–1535

152   Herodicus, being a trainer, and himself of a sickly constitution, by a combination of training and doctoring found out a way of torturing first and

chiefly himself, and secondly the rest of the world. By the invention of lin-
gering death; for he had a mortal disease; which he perpetually tended,
and, as recovery was out of the question, he passed his entire life as a vale-
tudinarian; he could do nothing but attend upon himself, and he was in
constant torment whenever he departed in anything from his usual regimen,
and so dying hard, by the help of science he struggled on to old age.
Utopia transl P K Marshall 1965 (New York: Washington Square Press) pp 59–60

Augustus De Morgan 1808–1871

153   Lagrange, in one of the later years of his life, imagined that he had overcome
the difficulty (of the parallel axiom). He went so far as to write a paper,
which he took with him to the Institute, and began to read it. But in
the first paragraph something struck him which he had not observed: he
muttered: '*Il faut que j'y songe encore*', and put the paper in his pocket.
[I must think about it again.]
Budget of Paradoxes London, 1872

Christian Morgenstern 1871–1914

154   *Lass die Moleküle rasen,*
Was sie auch zusammenknobeln!
Lass das Tüfteln, lass das Hobeln,
Heilig halte die Ekstasen!
Never mind mind molecular gyration,
Whatever be its chance creation!
Stop perfecting, quit cerebration
Give ecstasies your veneration.
Galgenlieder 1905. In G Stent *Paradoxes of Progress* 1978 (New York: W H Freeman)

Christopher Morley 1890–1957

155   An engineer gave me an ashtray
Made of a chunk of smelted bismuth.
The ore, when cooked,
Crystallises in cubes and terraces,
Condenses in sharp stairs and corners,
Like the ruins of a mimic Cuzco.
    O basic and everlasting geometry!
The cordillera itself
In the slack and purge of fire
Boils into right angles,
Takes conventional Inca pattern.
The greatest disorder on earth
Has the instinct of Perfect Form.
In Cady *Piezoelectricity*

[Lord] John Morley [of Blackburn] 1838–1923

156   The next great task of science is to create a religion for mankind.

William Morris 1834–1896

157   Science—we have loved her well, and followed her diligently, what will she
do? I fear she is too much in the pay of the counting-house, and the drill-
serjent, that she is too busy, and will for the present do nothing. Yet there

are matters which I should have thought easy for her; say, for example, teaching Manchester how to consume its town smoke, or Leeds how to get rid of its superfluous black dye without turning it into the river, which would be as much worth her attention as the production of the heaviest black silks, or the biggest of useless guns.
The Lesser Arts 1878

Samuel Finley Breese Morse 1791–1872

158   What hath God wrought.
[First message sent by him over the electric telegraph, from Washington to Baltimore, 24 May 1844]

Wolfgang Amadeus Mozart 1756–1791

159   [In 1782, about three piano concertos which he had just completed.] They are exactly between too hard and too easy. Very brilliant, pleasing to the ear, naturally without lapsing into emptiness, here and there only connoisseurs will find satisfaction, but in such a way that even non-connoisseurs must feel content without knowing why.
[Like a good lecture!]
Letter to his father, 28 December 1782. [K413, 414, 415]

Max Mueller 1823–1900

160   Language is the autobiography of the human mind.
In Nirad Chaudhuri *Scholar Extraordinary* 1974 (London: Chatto & Windus)

Herbert Joseph Muller 1905–

161   Another way of describing the revolution in physics is to say that the key nouns have been changed into verbs—to move, to act, to happen. What moves and acts, physicists do not care; 'matter' to them means 'to matter', to make a difference. But our language is still geared to express 'states of being', rather than processes. In this connection, also, the German language helps to explain German philosophy. The Germans have been especially prone to hypostatize their abstractions, identify the Rational and the Real, invent concepts comparable to Frankfurterness and Sauerkrautitude—for they capitalize all their nouns. And this may help to explain their present worship of the State.
Science and Criticism 1943 (New Haven, CT: Yale University Press) p 97

162   The great revolutionary thinkers are those who most violently wrenched traditional associations: Karl Marx was a philosophical Oscar Wilde, more scandalous because more sober.
Science and Criticism 1943 (New Haven, CT: Yale University Press) p 50

Hermann Joseph Muller 1891–1966

163   Death is an advantage to life.... Its advantage lies chiefly in giving ampler opportunities for the genes of the new generation to have their merits tested out...by clearing the way for fresh starts...
Science 1955 **121** 1

Robert Musil 1880–1942

164   [Austrian writer]
Wissen ist eine Sache der Einstellung, eine Leidenschaft, eigentlich eine

unerlaubte Einstellung. Denn der Zwang zum Wissen ist wie Trunksucht, wie Liebesverlangen, wie Mordlust, in dem sie einen Charakter aus dem Gleichgewicht wirft. Es stimmt doch gar nicht, dass der Wissenschaftler hinter der Wahrheit her ist. Sie is hinter ihm her. Er leidet unter ihr.

Knowledge is an attitude, a passion, actually an illicit attitude. For the compulsion to know is just like dipsomania, erotomania, homicidal mania, in producing a character that is out of balance. It is not true that the scientist goes after truth. It goes after him.

Der Mann ohne Eigenschaften 1930 (London: Picador) (also 1954 and 1979), 4 volumes (Peter Fischer, in his biography, *Licht und Leben*, of Max Delbrück, has chased this mysterious quotation, which appeared with the attribution to Kierkegaard in the first edition of this dictionary, to its probable origins. The search, involving many people, illustrates the proposition itself. J D Dunitz *Unverrichtete Dinge* 7 July 1989 (Zurich: ETH))

[Senator] Edmund Sixtus Muskie 1914–

165   [I want] a one-armed scientist...
[Who would not qualify his advice with 'on the other hand...']
Science 22 October 1976 **194** 390

Fridtjof Nansen 1861–1930

1   Man wants to know, and when he ceases to do so, he is no longer man.
[On the reason for polar explorations]

Napoleon I [Bonaparte] 1769–1821

2   The advance and perfecting of mathematics are closely joined to the prosperity of the nation.

3   They may say what they like; everything is organized matter.

Ogden Nash 1902–1971

4   Wind is caused by the trees waving their branches.

Gamel Abdel Nasser 1918–1970

5   The genius of you Americans is that you never make clear-cut stupid moves, only complicated stupid moves which make us wonder at the possibility that there may be something to them [which] we are missing.
[A text book of the game-theory view of politics as practised by the CIA]
In Miles Copeland *The Game of Nations* 1969 (London: Weidenfeld & Nicolson)

Sergei Gennadievich Nechaev 1847–1882

6   The revolutionist is a doomed man. He has no personal interests, no affairs, sentiments, attachments, property, not even a name of his own. Everything in him is absorbed by one exclusive interest, one thought, one passion—the revolution.

Joseph Needham 1900–

7   But Chinese civilization has the overpowering beauty of the wholly other, and only the wholly other can inspire the deepest love and the profoundest desire to learn.
The Grand Titration 1969 (London: Allen & Unwin)

8   Democracy might therefore almost in a sense be termed that practice of which science is the theory.
The Grand Titration 1969 (London: Allen & Unwin)

9   *Laboratorium est oratorium.*
The place where we do our scientific work is a place of prayer.
[A deduction from *laborare est orare*—to work is to pray]

10   I regarded the nature of biological organisation as a purely philosophical question, and excluded it from scientific biology. ...I had not seen the full significance of the analogous science of crystallography. I am glad to have an opportunity of cancelling what I then said.
Order and Life 1936 (Cambridge: Cambridge University Press)

11   It is mere verbiage...to talk about...'the hookworm's point of view' when its nervous system does not entitle it to have a point of view. Philosophers, on the contrary, are so entitled.
1936. *Moulds of Understanding* 1976 (London: Allen & Unwin)

Henry Needler 1685–1760

12   Who formed the curious texture of the eye,
      And cloath'd it with the various tunicles,
      And texture exquisite; with chrystal juice
      Supply'd it, to transmit the rays of light?
      A Poem to Prove the Certainty of a God in *Miscellaneous Correspondence* ed Benjamin Martin, London, 1759
      In M H Nicolson *Newton Demands the Muse* 1946 (Princeton, NJ: Princeton University Press)

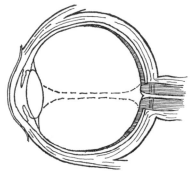

Jawaharlal Nehru 1889–1964

13   I fear that the spinning wheel is not stronger than the machine.
      [Gandhi and his followers promoted village industry rather than industrialization and their spinning wheel appears on the flag of India]

14   It is science alone that can solve the problems of hunger and poverty, insanitation and illiteracy, of superstition and deadening custom and tradition, of vast resources running to waste, of a rich country inhabited by starving people.... Who indeed could afford to ignore science today? At every turn we have to seek its aid.... The future belongs to science and to those who make friends with science.
      Proceedings of the National Institute of Sciences of India 1961 **27A** 564

Theodor Holm Nelson

15   This is the rock-solid principle on which the whole of the Corporation's [IBM's] Galaxy-wide success is founded...their fundamental design flaws are completely hidden by their superficial design flaws.
      Computer Lib. 1988 (London: Penguin). Reviewed in *The Guardian* 22 December 1988

Hermann Walther Nernst 1864–1941

16   [To students who disappointed him]
      For God's sake do not tell anyone that you studied under me.
      In *The Dictionary of Scientific Biography* ed C G Gillespie 1981 (New York: Scribner)

17   [On examinations]
      Das Wissen ist der Tod der Forschung.
      Knowledge is the death of research.
      In *The Dictionary of Scientific Biography* ed C G Gillespie 1981 (New York: Scribner)

Johann von Neumann 1903–1957

18   In mathematics you don't understand things. You just get used to them.
      In G Zukav *The Dancing Wu Li Masters* (London: Hutchinson)

[Cardinal] John Henry Newman 1801–1890

19   Living movements do not come out of committees.

[Sir] Isaac Newton 1642–1727

20   Are not gross bodies and light convertible into one another; and may not
      bodies receive much of their activity from the particles of light which enter
      into their composition? The changing of bodies into light, and light into
      bodies, is very conformable to the course of Nature, which seems delighted
      with transmutations.
      Opticks 1704, Query 30

21   I feign no hypotheses (*hypotheses non fingo*), for whatever is not deduced
      from the phenomena is to be called a hypothesis, and hypotheses, whether
      metaphysical or physical, whether of occult qualities or mechanical have
      no place in experimental philosophy. In this philosophy particular proposi-
      tions are inferred from the phenomena and afterwards rendered general by
      induction.
      Scholium to *Philosophiae Naturalis Principia Mathematica*

22   I intend, to be no further solicitous about matters of Philosophy; and there-
      fore I hope you will not take it ill, if you find me never doing anything more
      in that kind.
      [Letter to Oldenburg, the Secretary of the Royal Society]
      Opticks 1704

23   I know not what I may appear to the world, but to myself I seem to have
      been only like a boy playing on the sea-shore, and diverting myself in now
      and then finding a smoother pebble or a prettier shell than ordinary, whilst
      the great ocean of truth lay all undiscovered before me.
      In D Brewster *Memoirs of Newton* 1855 vol 2 Chapter 27

24   Physics, beware of metaphysics.

25   Whence is it that nature does nothing in vain; and whence arises all that
      order and beauty which we see in the world?
      Opticks 1704

26   The latest authors, like the most ancient, strove to subordinate the phe-
      nomena of nature to the laws of mathematics.
      I S Dmitriev

27   It seems probable to me that God in the beginning formed matter in solid,
      massy, hard, impenetrable, moveable particles, of such sizes and figures
      and with such other properties and in such proportion to space, as most
      conduced to the end for which he formed them; and that these primitive
      particles being solids are incomparably harder than any porous bodies com-
      pounded of them, even so very hard as never to wear or break in pieces.
      Opticks 1704, IV.260

28   Should we not suppose that in the formation of a crystal the particles are
      not only established in rows and columns set in regular figures, but also

by means of some polar property have turned identical sides in identical directions?
R V Bogdanov

29   I do not define time, space, place and motion, as being well known to all.
[See: first entry under ANON, about the *Polish Encyclopaedia*]
Scholium 8 in *Principia Mathematica*

30   Our business is with the causes of sensible effects.
In *Physics Bulletin* September 1977

31   *Numero pondere et mensura Deus omnia condidit.*
God created everything by number, weight and measure.
Dedication in the autograph book of Ferenc Páriz Pápai, 11 September 1722
[*Wisdom of Solomon* 11:20]

32   [Newton, having determined "the motions of the planets, the comets, the Moon and the sea", was unfortunately unable to determine the remaining structure of the world from the same propositions because:]
I suspect that they may all depend upon certain forces by which the particles of the bodies, by some causes hitherto unknown, are either mutually impelled towards one another, and cohere in regular figures, or are repelled and receed from one another. These forces being unknown, philosophers have hitherto attempted the search of Nature in vain; but I hope the principles laid down will afford some light either to this or some truer method of philosophy.
Preface to the *Principia.... Nature* 22 October 1987 **329**, 772

33   [Asked how he made his discoveries] By always thinking unto them. I keep the subject constantly before me and wait till the first dawnings open little by little into the full light.
In *Nature* 4 September 1965

Norman Nicholson 1914–

34   The toadstool towers infest the shore:
Stink-horns that propagate and spore
Wherever the wind blows.
Scafell looks down from the bracken band,
And sees hell in a grain of sand,
And feels the canker itch between his toes.
  This is a land where dirt is clean,
And poison pasture, quick and green,
And storm sky, bright and bare;
Where sewers flow with milk, and meat
Is carved up for the fire to eat,
And children suffocate in God's fresh air.
[On the leak of radioactive iodine and polonium from Windscale]
Windscale in *A Local Habitation* 1972 (London: Faber & Faber)

35   The furthest stars recede
Faster than the earth itself to our need.
  For far beyond the furthest, where
Light is snatched backward, no
Star leaves echo or shadow

To prove it has ever been there.
  And if the universe
Reversed and showed
The colour of its money;
If now unobservable light
Flowed inward, and the skies snowed
A blizzard of galaxies,
The lens of night would burn
Brighter than the focussed sun,
And man turn blinded
With white-hot darkness in his eyes.
The Expanding Universe in *The Pot Geranium* 1954 (London: Faber & Faber)

Marjorie Hope Nicolson 1894–

36   The language of poetry and science was no longer one when the world was no longer one.
In *Encyclopaedia of Poetry & Poetics* 1965 (Princeton, NJ: Princeton University Press)

Friedrich Nietzsche 1844–1900

37   ...—it is all over with priests and gods when man becomes scientific. Moral: science is the forbidden as such—it alone is forbidden. Science is the first sin, seed of all sin, the original sin. This alone is morality. 'Thou shalt not know'—the rest follows.
Antichrist Chapter 48

38   *Glaubt ihr denn, dass die Wissenschaften entstanden und gross geworden wären, wenn ihnen nicht Zauberer, Alchimisten, Astrologen und Hexen vorangelaufen wären als die, welche erst Durst, Hunger und Wohlgeschmack an verborgenen und verbotenen Mächten schaffen mussten?*
Do you believe then that the sciences would ever have arisen and become great if there had not beforehand been magicians, alchemists, astrologers and wizards, who thirsted and hungered after abscondite and forbidden powers?
Die fröhliche Wissenschaft 1886, IV

39   'Then art thou perhaps an expert on the leech?' asked Zarathustra 'and thou investigatest the leech to its ultimate basis, thou conscientious one?' 'O Zarathustra' answered the trodden one, 'that would be something immense: how could I presume to do so! That, however, of which I am master and knower, is the brain of the leech:—that is my world!'
Neurobiology of the Leech 1981 (New York: Cold Spring Harbour). *Also Sprach Zarathustra* in *Scientific American* May 1982

40   *Kein Sieger glaubt an den Zufall.*
No victor believes in chance.
Die fröhliche Wissenschaft, 258, 1886

41   *Wir müssen Physiker sein, um, in jenem Sinne, Schöpfer sein zu können—während bisher alle Wertschätzungen und Ideale auf Unkenntnis der Physik oder im Widerspruch mit ihr aufgebaut waren.*
We must be physicists in order, in that sense, to be creative since so far

codes of values and ideals have been constructed in ignorance of physics or even in contradiction to physics.
Die fröhliche Wissenschaft, 335, 1882–1887

Florence Nightingale 1820–1910

42   [Of her] Her statistics were more than a study, they were indeed her religion. For her Quetelet was the hero as scientist, and the presentation copy of his *Physique Sociale* is annotated by her on every page. Florence Nightingale believed—and in all the actions of her life acted upon that belief—that the administrator could only be successful if he were guided by statistical knowledge. The legislator—to say nothing of the politician—too often failed for want of this knowledge. Nay, she went further; she held that the universe—including human communities—was evolving in accordance with a divine plan; that it was man's business to endeavour to understand this plan and guide his actions in sympathy with it. But to understand God's thoughts, she held we must study statistics, for these are the measure of His purpose. Thus the study of statistics was for her a religious duty.
In Karl Pearson *The Life, Letters and Labours of Francis Galton* vol 2, 1924 (London: Cambridge University Press). *Isis* **8** 186

43   The true foundation of theology is to ascertain the character of God. It is by the art of Statistics that law in the social sphere can be ascertained and codified, and certain aspects of the character of God thereby revealed. The study of statistics is thus a religious service.
In F N David *Games, God and Gambling* 1962 (London: Griffin)

Yasunori Nishijima

44   We don't need more creativity in Japan. One creative mind in a thousand is perfectly sufficient. Any more and the country would collapse in chaos.
The Guardian 14 December 1987

Richard Milhous Nixon 1913–

45   Listen, I don't know anything about polygraphs and I don't know how accurate they are, but I know they'll scare the hell out of people.
Oval Office Tape, 14/7/1971. In *Nature* 23 February 1984 **307** 682

Novalis [Friedrich Leopold von Hardenberg] 1772–1801

46   *Die Mathematik ist das Leben der Götter.*
Mathematics is the Life of the Gods.

47   [Water is] sensitive chaos.

Alfred Noyes 1880–1958

48   ...a flock of crazy prophets, that by staring at a crystal can fill it with more fancies than there are herrings in the sea.
Wizards

William of Occam 1300–1349

1   [From the village of Ockham in Surrey]
Frustra fit per plura, quod fieri potest per pauciora.
or: *Essentia non sunt multiplicanda praeter necessitatem.*
It is vain to do with more what can be done with less.
or: Entities are not to be multiplied beyond necessity.
Occam's Razor

Charles Kay Ogden 1889–1957

2   The belief that words have a meaning of their own account is a relic of
primitive word magic, and it is still a part of the air we breathe in nearly
every discussion.
The Meaning of Meaning 1923 (London: Kegan Paul)

Charles Kay Ogden 1889–1957 and **Ivor Armstrong Richards** 1893–1979

3   The gostak distims the doshes.
[There is an excellent science-fiction story on this theme by Miles J Brener which
appeared in *Amazing Stories* in March 1930]
The Meaning of Meaning 1923 (London: Kegan Paul)

Sorai Ogyu 1666–1728

4   Mathematicians boast of their exacting achievements, but in reality they
are absorbed in mental acrobatics and contribute nothing to society.
Complete Works on Japan's Philosophical Thought 1956 (Tokyo: Heibonsha)

Henry Oldenburg ca 1626–1678

5   I acknowledge that the jealousy about the first authors of experiments
which you speak of is not groundless; and therefore offer myself to register
all those you, or any person, shall please to communicate as new, with
that fidelity, which both the honour of my relation to the Royal Society
(which is highly concerned in such experiments) and my own inclinations,
do strongly oblige me to.
[The first editor of the *Philosophical Transactions of the Royal Society* writing to
Robert Boyle]
Isis 1940 **31** 321

Bernard More Oliver 1916–

6   It is time that science, having destroyed the religious basis for morality,
accepted the obligation to provide a new and rational basis for human
behaviour—a code of ethics concerned with man's needs on earth, not his
rewards in heaven.
Towards a New Morality in *IEEE Spectrum* 1972 **9** 52

Omar Khayyam ca 1050–ca 1123

7   Myself when young did eagerly frequent
Doctor and Saint, and heard great Arguement
About it and about: but everymore
Came out by the same door as in I went.
The Rubiyat 1859, transl Edward Fitzgerald, I.28

8   Khayyam, who stitched the tents of science...
[On his own poetical name which signifies a tentmaker]
In E Fitzgerald *The Rubiyat of Omar Khayyam* 1905. Introduction

9   For in and out, above, about, below,
'Tis nothing but a Magic Shadow-show,
Played in a Box whose Candle is the Sun,
Round which we Phantom Figures come and go.
In E Fitzgerald *The Rubiyat of Omar Khayyam* 1905. Quatrain XLVI

Eugene O'Neill 1888–1953

10   Happiness hates the timid! So does science!
Strange Interlude, 4, 1928

Julius Robert Oppenheimer 1904–1967

11   In some sort of crude sense which no vulgarity, no humor, no overstatement
can quite extinguish, the physicists have known sin; and this is a knowledge
which they cannot lose.
[Lecture at MIT, 25 November 1947]
Physics in the Contemporary World (Cambridge, MA: Harvard University Press)

12   There are children playing in the street who could solve some of my top
problems in physics, because they have modes of sensory perception that I
lost long ago.

13   There floated through my mind a line from the *Bhagavad-Gita* in which
Krishna is trying to persuade the Prince that he should do his duty: 'I am
become death, the shatterer of worlds'. I think we all had this feeling more
or less.
[On 16 July 1945, at the test of the first atomic bomb—Trinity. The previous lines of
the *Bhagavad-Gita* are:
If the radiance of a thousand suns
Were to burst into the sky,
That would be like
The splendour of the Mighty One]
In N P Davis *Lawrence and Oppenheimer* 1969 (London: Cape)

14   When you see something that is technically sweet, you go ahead and do
it and you argue about what to do about it only after you have had your
technical success.
In R W Reid *Tongues of Conscience: Weapons Research and the Scientist's Dilemma*
1969 (New York: Walker)

15   Today, it is not only that our kings do not know mathematics, but our
philosophers do not know mathematics and—to go a step further—our
mathematicians do not know mathematics.
The Tree of Knowledge in *Harper's* 1958 **217** 55–7

16   We live today in a world in which poets and historians and men of affairs are
proud that they wouldn't even begin to consider thinking about learning
anything of science, regarding it as the far end of a tunnel too long for any
wise man to put his head into.
The Open Mind 1955 (New York: Simon & Schuster)

17   The scientist is not responsible for the laws of nature, but it is a scientist's job to find out how these laws operate. It is the scientist's job to

find ways in which these laws can serve the human will. However, it is not the scientist's job to determine whether a hydrogen bomb should be used. This responsibility rests with the American people and their chosen representatives.

In L Wolpert and A Richards (ed) *A Passion for Science* 1988 (Oxford: Oxford University Press)

Martin Oppenheimer 1930–

18   Today's city is the most vulnerable social structure ever conceived by man.
Urban Guerilla (London: Quadrangle)

José Ortega y Gasset 1883–1955

19   Contemporary science, with its system and methods, can put blockheads (*tontos*) to good use.
Obras Completas. Revista de Occidente Madrid 1958, vol 6

20   In the disturbances caused by scarcity of food, the mob goes in search of bread, and the means it employs is generally to wreck the bakeries. This may serve as a symbol of the attitude adopted, on a greater and more complicated scale, by the masses of today towards the civilisation by which they are supported.
The Revolt of the Masses 1961 (London: Allen & Unwin)

George Orwell 1903–1950

21   In a way it is even humiliating to watch...miners working. It raises in you a momentary doubt about your own status as an 'intellectual' and a superior person generally. For it is brought home to you, at least while you are watching, that it is only because miners sweat their guts out that superior persons can remain superior. You and I and the editors of the *Times Literary Supplement,* and the Poets and the Archbishop of Canterbury and Comrade X, author of *Marxism for Infants*—all of us really owe the comparative decency of our lives to poor drudges underground...with their throats full of...dust, driving their shovels forward with arms and belly muscles of steel.
Down the Mines in *The Road to Wigan Pier* 1937 (London: Secker & Warburg)

22   Who controls the past controls the future. Who controls the present controls the past.
[The motto of the Ministry of Truth for which the Senate House building of the University of London was the model.]
Nineteen Eighty-four 1949 (London: Secker & Warburg)

[Sir] William Osler 1849–1919

23   In science the credit goes to the man who convinces the world, not to the man to whom the idea first occurs.
Probably: *Aequanimitas* with other Addresses in *Books and Men*

W J Osterhout

24   You geneticists may know something about the hereditary mechanisms that distinguish a red-eyed from a white-eyed fruit fly but you haven't the slightest inkling about the hereditary mechanism that distinguishes fruit flies from elephants.
1925. In P C Mangelsdorf *Corn. Its origin, Evolution and Improvement* 1974 (Cambridge, MA: Belknap/Harvard University Press)

Wilhelm Ostwald 1853–1932

25   We, or rather the Germanic race, have discovered the factor of organisation.
Other peoples still live under the regime of individualism, whereas we live
under the regime of organisation.
In J Labadie (ed) *L'Allemagne, a-t-elle le Secret de L'organisation?* 1916, Paris

Richard Overton 1631–1664

26   [The leveller] Form is always the form of matter, and matter is the matter
of form. Each of them cannot exist by itself alone but only in unity with
the other, and only in unity do they form a thing.

Ovid 43 BC–17 AD

27   *Nihil est toto, quo perstet, in orbe*
Cuncta fluunt, omnisque vagans formatur imago
Ipsa quoque odsidue labuntur tempora motu.
There is nothing in the whole world which is permanent
Everything flows onward; all things are brought into being with a changing
nature;
The ages themselves glide by in constant movement.
Metamorphoses XV, i, 177

28   *Desinet ante dies et in alto Phoebus anhelos aequore tinquiet equos, quam*
consequar omnia verbis in species translata novas.
Evening has overtaken me, and the sun has dipped below the horizon of
the Ocean, yet I have not had time to tell you of all the things that have
evolved into new forms.
Metamorphoses

Axel Gustafsson [Count] Oxenstierna 1583–1654

29   *Nescis, mi fili, quantilla ratione mundus regatur.*
You don't know, my dear boy, with what little reason the world is governed.
Letter to his son, 1648

Juho Kusti Paasikivi 1870–1965

1   [Former President of Finland]
      Kaiken viisauden alku on tosiasiain tunnustaminen.
      Recognition of realities is the beginning of all wisdom.
      On his tomb in Helsinki

Denys Lionel Page

2   [On the bureaucracy of Pylos as evinced by the inscriptions in Linear B]
      One would suppose that not a seed could be sown, not a gram of bronze
      worked, not a cloth woven, not a goat reared or a hog fattened, without
      the filling of a form in the Royal Palace; such is the impression made by
      only part of the files for a single year.
      History and the Homeric Iliad, Sather Classical Lectures, XXXI, Berkeley, California,
      1959 (Berkeley, CA: University of California Press)

[Sir] James Paget 1814–1899

3   You will find that fatigue has a larger share in the promotion and trans-
      mission of disease than any other single condition you can name.
      Science in War 1940 (London: Penguin)

William Paley 1743–1805

4   I take my stand in human anatomy...the necessity, in each particular case,
      of an intelligent designing mind for the contriving and determining of the
      forms which organised bodies bear.
      [Illustration of Paley's watch from Bernard Nieuwentyt (1654–1718), a disciple of
      Descartes. The germ of the idea was earlier in Cicero: *De Natura Deorum*, ii, 34]
      Natural Theology

Bernard Palissy 1510–1589

5   You must know that, in order to manage well a kiln full of pottery, even
      when it is glazed, you must control the fire by so careful a philosophy that
      there would be no spirit however noble which would not be much tried
      and often disappointed. As to the manner of filling your kiln, a singular
      geometry is needed.... The arts for which compass, ruler, numbers, weights
      and measures are needed should not be called mechanics.
      [Palissy was the potter who found science for himself]
      L'Art de Terre in *Oeuvres Complètes* Paris, 1884

George Paloczi-Horvath 1908–1973

6   Students of Soviet affairs know how difficult it is to fortell the Soviet past.
      Khrushchev 1960 (Boston, MA: Little, Brown)

Paracelsus [Philippus Aureolus Theophrastus Bombastus von Hohen-heim] 1493–1541

7   To each elemental being the element in which it lives is transparent, invis-
      ible and respirable, as the atmosphere to ourselves.
      In F Hartmann *The Life of...Paracelsus* 1896, London, 2nd edn

8   What is accomplished with fire is alchemy, whether in the furnace or the
      kitchen stove.
      In J Bronowski *The Ascent of Man* 1975 (London: BBC)

9   I have not patched up these books after the fashion of others from Hip-
     pocrates, Galen or anyone else; but by experience, the great teacher, I have
     composed them.
     British Museum exhibition of scientific books

10   The true use of chemistry is not to make gold but to prepare medicines.
     British Museum exhibition of scientific books

Vilfredo Pareto 1848–1923

11   Give me fruitful error any time, full of seeds, bursting with its own correc-
     tions. You can keep your sterile truth for yourself.
     [Comment on Kepler]

12   In a dispute between two chemists there is a judge; Experience. In a dispute
     between a Moslem and a Christian, who is the judge? Nobody.
     The Mind and Society

[Sir] Alan Sterling Parkes 1900–

13   ...no woman should be kept on the Pill for 20 years until, in fact, a sufficient
     number have been kept on the Pill for 20 years.
     Nature 11 April 1970 **226** 187

Cyril Northcote Parkinson 1909–

14   Work expands so as to fill the time available for its completion (Parkinson's
     First Law).
     Parkinson's Law 1957 (London: Murray)

15   Expenditure rises to meet income (Parkinson's Second Law).
     In Laws and Outlaws 1962 (London: Penguin)

16   Expansion means complexity, and complexity decay (Parkinson's Third
     Law).
     In Laws and Outlaws 1962 (London: Penguin)

17   The Law of Triviality:
     Briefly stated, it means that the time spent on any item of the agenda will
     be in inverse proportion to the sum involved.
     Parkinson's Law 1957 (London: Murray)

Talcott Parsons 1902–1979

18   Science is intimately integrated with the whole social structure and cul-
     tural tradition. They mutually support one another—only in certain types
     of society can science flourish, and conversely without a continuous and
     healthy development and application of science such a society cannot func-
     tion properly.
     The Social System 1951 (New York: Free Press) Chapter VIII

Blaise Pascal 1623–1662

19   [I feel] engulfed in the infinite immensity of spaces whereof I know nothing,
     and which know nothing of me, I am terrified.... The eternal silence of
     these infinite spaces alarms me.
     Pensées 1657

20   I have made this letter longer than usual because I lack the time to make
     it shorter.
     Lettres Provinciales, XVI

Louis Pasteur 1822–1895

21   *Dans les champs de l'observation, l'hasard ne favorise que les esprits*
préparés.
In the field of observation, chance only favours those minds which have
been prepared.
Encyclopaedia Britannica 1911, 11th edn, vol 20

22   *Il faut de toute necessité que des actions dissymetriques president pendant*
la vie à l'élaboration des vrais principes immédiats naturels dissymetriques.
Quelle peut être la nature de ces actions dissymetriques? Je pense, quant
à moi, qu'elles sont d'ordre cosmique. L'univers est un énsemble dis-
symetrique et je suis persuadé que la vie, telle qu'elle se manifeste à nous,
est fonction de la dissymetrie de l'univers ou des consequences qu'elle en-
traîne. L'univers est dissymetrique.
It is inescapable that asymmetric forces must be operative during the synthesis of the first asymmetric natural products. What might these forces
be? I, for my part, think that they are cosmological. The universe is asymmetric and I am persuaded that life, as it is known to us, is a direct result of
the asymmetry of the universe or of its indirect consequences. The universe
is asymmetric.
Comptes Rendus de l'Académie des Sciences 1 June 1874. Reprinted *Oeuvres* 1, 361.
In J B S Haldane *Nature* 1960 **185** 87

23   There does not exist a category of science to which one can give the name
applied science. There are science and the applications of science, bound
together as the fruit of the tree which bears it.
Pourquoi la France n'a pas trouvé d'hommes supérieurs au moment du peril in *Revue*
Scientifique 1871

24   Unfortunate those scientists who have only clear thoughts in their heads!
In a lecture by A Eschenmoser, 1976

25   *La science n'a pas de patric.*
Science owns no fatherland.
Discours du 14 Novembre 1888 pour l'inauguration de l'Institut Pasteur

Patanjali ca 300–100 BC

26   One goes to the potter for pots, but not to the grammarian for words.
Language is already there among the people.
Mahabhasya. Science and Technology in Indian Culture, NISTADS, New Delhi, 1984.

Howard Pattee

27   It is better to be pessimistic and find out you are wrong, than to be opti-
mistic and find out nothing.
In S W Fox (ed) *The Origins of Prebiological Systems and their Molecular Matrices*
1963.

Wolfgang Pauli 1900–1958

28   *Ich habe nichts dagegen wenn Sie langsam denken, Herr Doktor, aber ich*
habe etwas dagegen wenn Sie rascher publizieren als denken.
I don't mind your thinking slowly: I mind your publishing faster than you
think.
Attributed (from H Coblaus)

29   I refuse to believe that God is a weak left-hander.

Ivan Petrovich Pavlov 1849–1936

30   What can I wish to the youth of my country who devote themselves to science.... Thirdly, passion. Remember that science demands from a man all his life. If you had two lives that would not be enough for you. Be passionate in your work and in your searching.
Bequest to Academic Youth 1936

Edward John Payne 1844–1904

31   So far as concerns the New World, the above facts tend to support the general proposition, to which the history of advancement in the Old World suggests no exception, that nothing worthy of the name of civilisation has ever been founded on any other agricultural basis than [that of] the cereals.... Cereal agriculture, alone among the forms of food-production, taxes, recompenses and stimulates labour and ingenuity in an equal degree.
History of the New World called America, 1892

Karl Pearson 1857–1936

32   Modern science, as training the mind to an exact and impartial analysis of facts, is an education specially fitted to promote citizenship.
The Grammar of Science 1911 (London: A & C Black)

33   The right to live does not connote the right of each man to reproduce his kind. ...As we lessen the stringency of natural selection, and more and more of the weaklings and the unfit survive, we must increase the standard, mental and physical, of parentage.
Darwinism, Medical Progress and Parentage 1912, University of London, University College Eugenics Laboratory, 2nd edn

34   The unity of all science consists alone in its method, not in its material.
The Grammar of Science 1911 (London: A & C Black)

Benjamin Peirce 1809–1880

35   Mathematics is the science which draws necessary conclusions.
[Memoir read before the National Academy of Sciences in Washington, 1870]
American Journal of Mathematics 1881 **4** 97

Pelagius [Morgan] ca 360–ca 420

36   *Si necessitatis est, peccatum non est; si voluntatis, vitari potest.*
If it is a necessity, then it is not a sin; if it is optional, then it can be avoided.
[The founder of the Pelagian heresy]

Pericles died 429 BC

37   Although only a few may originate a policy, we are all able to judge it.
Funeral Oration 431 BC. *Thucydides* II, 41

Jean Perrin 1870–1942

38   *Habitués par le travail de laboratoire aux prévisions lucides, nous apercevons clairement ce que ne devinent pas encore les ignorants; et je range dans cette catégorie des ignorants certains hommes cultivés mais complètement*

inconscients de la Science et de ses prodigieuses possibilités, et qui, par une aberration inconcevable en présence de ce que nous avons déjà vu, pensent que l'avenir ressemblera nécessairement au passé, en sorte qu'il y aurait toujours des guerres, toujours des pauvres, toujours des esclaves!

Accustomed as we are, by laboratory work, to clear predictions, we see clearly what the more ignorant still have not realised; and I put into this category of ignorant certain men who are cultivated, but are completely unaware of science and its enormous potential, and who, by an aberration, unimaginable in the light of what we have already seen, think that the future will always have to be like the past and conclude that there will always be wars, poverty and slavery.

La Science et l'Espérence 1948 (Paris: Presses Universitaires de France

Max Ferdinand Perutz 1914–

39   People sneer at Jim's book because they say that all he did in Cambridge was play tennis and chase girls. But there was a serious point to that. I sometimes envied Jim. My own problem took thousands of hours of hard work, measurements, calculations. I often thought that there must be, if only I could see it, an elegant solution. There wasn't any. For Jim's there was an elegant solution, which is what I admired. He found it partly because he never made the mistake of confusing hard work with hard thinking; he always refused to substitute the one for the other. Of course he had time for tennis and girls.

Horace Judson, *The New Yorker*, 27 November 1978

40   The author, like Dante, descends from the heavens of pure crystalline order, through a purgatory of various disorders, to the depths of complete infernal chaos.

1966. Foreword to the English edition of B K Vainshtein's book *Diffraction from Chain Molecules*

Laurence Johnston Peter 1919–

41   The Peter Principle:
In a hierarchy every employee tends to rise to his level of incompetence. [Whence two sub-principles]—In time, every post tends to be occupied by an employee who is incompetent to carry out its duties.—Work is accomplished by those employees who have not yet reached their level of incompetence.

The Peter Principle 1969 (New York: Morrow)

Petronius [Gaius Petronius Arbiter] 1st Century AD

42   We trained hard but it seemed that every time we were beginning to form teams we would be reorganised. I was to learn later in life that we tend to meet every situation in life by reorganising, and a wonderful method it can be for creating the illusion of progress while producing confusion, inefficiency and demoralisation.

In G Porter, President of the Royal Society, Anniversary Address, 30 November 1989

[Sir] William Petty 1623–1687

43   Nor do I doubt if the most formidable armies ever heere upon earth is a sort of soldiers who for their smallness are not visible.
[On microbes, 1640]
The Petty Papers 1927 (London: Constable)

Physics Review Letters

44   Scientific discoveries are not the proper subject for newspaper scoops and all media of mass communication should have equal opportunity for simultaneous access to the information. In the future we may reject papers whose main contents have been published previously in the daily press.

[Editorial by Professor Samuel A Goudsmit, 1 January 1960. Since Professor Goudsmit's retirement as editor 'the publicity-hungry high-energy physicists have succeeded in getting this old policy rescinded']

Jean Piaget 1896–1980

45   In short, the notion of structure is comprised of three key ideas: the idea of wholeness, the idea of transformation, and the idea of self-regulation.

Structuralism transl C Maschler, 1971 (London: Routledge & Kegan Paul)
Le Structuralisme 1970 (Paris: Paris University Press)

Pablo [Ruiz y] Picasso 1881–1973

46   The against comes before the for.

47   *Je ne cherche pas; je trouve.*
I do not search; I find.

Étude de Femme in P Oster Nouveau Dictionnaire de Citations Franaises 1970 (Paris: Librairie Hachette, Tchou Editeurs)

Gerard Piel 1915–

48   [Editor of *Scientific American*]
The most remarkable discovery made by scientists is science itself.

In J Bronowski *Magic, Science and Civilisation* 1978 (New York: Columbia University Press)

Charles Santiago Sanders Pierce 1839–1914

49   The essence of belief is the establishment of a habit.

Illustrations of the Logic of Science, II in *Popular Science Monthly, NY* January 1878

50   [Science] advances by leaps; and the impulse for each leap is either some new observational resource, or some novel way of reasoning about the observations. Such novel way of reasoning might, perhaps, be considered as a new observational means, since it draws attention to relations between facts which would previously have been passed by unperceived.

ca 1896. *Collected Papers*, I.1.109. Ed C Hartshorne *et al*, 1931, Cambridge, MA

[Sir] Alfred Brian Pippard 1920–

51   The value of a formalism lies not only in the range of problems to which it can be successfully applied, but equally in the degree to which it encourages physical intuition in guessing the solution of intractable problems.

Physics Bulletin 1969 **20** 455

Norman Wingate Pirie 1907–

52   My teacher, Hopkins, often commented on the craving for certainty that led so many physicists into mysticism or into the Church and similar organisations.... Faith seems to be an occupational hazard for physicists.

Penguin New Biology 1954 **16** 44

Robert Pirsig 1929–

53   The way to solve the conflict between human values and technological needs
is not to run away from technology, that's impossible. The way to resolve
the conflict is to break down the barriers of dualistic thought that prevent
a real understanding of what technology is—not an exploitation of nature,
but a fusion of nature and the human spirit into a new kind of creation
that transcends both.
Zen and the Art of Motorcycle Maintenance 1974 (London: Bodley Head)

54   The Buddha, the Godhead, resides quite as comfortably in the circuits of
a digital computer or the gears of a cycle transmission as he does at the
top of a mountain or in the petals of a flower.
Zen and the Art of Motorcycle Maintenance 1974 (London: Bodley Head)

55   A motorcycle functions entirely in accordance with the laws of reason, and
a study of the art of motorcycle maintenance is really a miniature study of
the art of rationality itself.
Zen and the Art of Motorcycle Maintenance 1974 (London: Bodley Head)

[Pope] Pius IX 1792–1878

56   Syllabus of the principal errors of our time...12. The decrees of the Apostolic See and of the Roman congregation impede the true progress of science.
[21 December 1863]
In *Dogmatic Canons and Decrees* 1912 (Old Greenwich, CT: Devin–Adair)

[Pope] Pius XII 1876–1958

57   The Church welcomes technological progress and receives it with love, for
it is an indubitable fact that technological progress comes from God and,
therefore, can and must lead to Him.
[Christmas Message, 1953]

58   One Galileo in two thousand years is enough.
[On being asked to proscribe the works of Teilhard de Chardin]
Attributed. See Stafford Beer *Platform for Change* 1975 (Chichester: Wiley)

Max Planck 1858–1947

59   An important scientific innovation rarely makes its way by gradually win-
ning over and converting its opponents: it rarely happens that Saul becomes
Paul. What does happen is that its opponents gradually die out, and that
the growing generation is familiarised with the ideas from the beginning.
In G Holton *Thematic Origins of Scientific Thought* 1973 (Cambridge, MA: Harvard
University Press)
Scientific Autobiography 1949 (New York: Philosophical Library)

Plato ca 429–347 BC

60   He who can properly define and divide is to be considered a god.
In Francis Bacon *Novum Organum* 1620, Book II, 26

61   The ludicrous state of solid geometry made me pass over this branch.
The Republic VII, 528

62   We must endeavour to persuade the principal men of our State to go and
learn arithmetic, not as amateurs,...but for the sake of military use...
The Republic VII, 525

63   Let no one ignorant of geometry enter my door.
       In John Tzetzes (ca 1110–ca 1180) *Chiliad* 8, 972

64   He is unworthy of the name of man who is ignorant of the fact that the
       diagonal of a square is incommensurable with its side.
       In Sophie Germain *Mémoire sur les surfaces élastiques*

65   *Socrates*: Shall we set down astronomy among the subjects of study?
       Glaucon: I think so, to know something about the seasons, the months and
       the years is of use for military purposes, as well as for agriculture and for
       navigation.
       Socrates: It amuses me to see how afraid you are, lest the common herd of
       people should accuse you of recommending useless studies.
       The Republic VII, 527

66   *Theaetetus*: Science is sensation.
       Socrates: You give an opinion that cannot be despised, since it was Pro-
       tagoras's. Yet he expressed it in another way, by saying that man was the
       measure of all things.
       Theaetetus 152

67   [The Earth is like] one of those balls made of twelve pieces of skin.
       Theatetus. Problem 52 of Phaedo, 110b

68   [Of the five Platonic solids]
       So their combinations with themselves and with each other give rise to
       endless complexities, which anyone who is to give a likely account of reality
       must survey.
       The Timaeus, Conclusion

John Rader Platt 1918–

69   Strong inference consists in applying the following steps to a problem in
       science, formally and explicitly and regularly: (1) devising alternative hy-
       potheses; (2) devising a crucial experiment (or several of them) with alter-
       native possible outcomes, each of which will, as nearly as possible, exclude
       one or more of the hypotheses; (3) carrying out the experiment so as to
       get a clean result; and recycling the procedure, making subhypotheses or
       sequential hypotheses to define the possibilities that remain; and so on.
       In C H Waddington *The Tools of Thought* 1977 (London: Cape)

Georgi Valentinovich Plekhanov 1856–1918

70   Bourgeois scientists make sure that their theories are not dangerous to God
       or to capital.
       Karl Marx 1903

Pliny [Gaius Plinius Secundus, the Elder] 23–79

71   *Quare sexangulis nascatur lateribus non facile ratio invenire potest; eo*
       magis quod neque mucronis eadem species est, et ita absolutus est laterum
       laevor, ut nulla id arte possit aequari.
       [Of Quartz] Why it is formed with hexagonal faces cannot be readily ex-
       plained; and any explanation is complicated by the fact that, on the one
       hand, its terminal points are not symmetrical and that, on the other hand,
       its faces are so perfectly smooth that no craftsmanship could achieve the
       same effect.
       Natural History Book 37.9.

Plutarch ca 46–ca 127

72   But what most of all afflicted Marcellus was the death of Archimedes. For
it chanced that he was by himself, working out some problem with the aid
of a diagram, and having fixed his thoughts and his eyes as well upon the
matter of his study, he was not aware of the incursion of the Romans, or of
the capture of the city. Suddenly a soldier came upon him and ordered him
to go with him to Marcellus. This Archimedes refused to do until he had
worked out his problem and established his demonstration, whereupon the
soldier flew into a passion, drew his sword, and dispatched him. However,
it is generally agreed that Marcellus was afflicted at his death, and turned
away from his slayer as from a polluted person, and sought out the kindred
of Archimedes and paid them honour.
Lives transl J & W Langhorne 1876 (London: Chatto)

73   Plato said that God geometrises continually.
Convivialum Disputationum 8, 2

74   You know, of course, that Lycurgus expelled arithmetical proportion from
Lacedaemon, because of its democratic and rabble-rousing character. He
introduced geometric proportion,...
Moralia Loeb edn, vol IX

75   Is their opinion true who think that he ascribed a dodecahedron to the
globe, when he says that God made use of its bases and the obtuseness
of its angles, avoiding all rectitude. It is flexible and by circumtension,
like globes made of twelve skins, it becomes circular and comprehensive
[like the current football]. For it has twenty solid angles, each of which
is contained by three obtuse planes, and each of those contains one and
the fifth part of a right angle. Now it is made up of twelve equilateral and
equiangular quinquangles (or pentagons), each of which consists of thirty of
the first scalene triangles. Therefore, it seems reasonable to resemble both
the Zodiac and the year being divided into the same number of parts as
these.
Quaestiones Platonicae, 5.1, 1003C (transl by R Brown), in *Plutarch's Morals* ed W
W Goodwin (1870), vol 5

Po Chu-i 772–846

76   'Those who speak know nothing;
Those who know are silent'.
These words, as I am told,
Were spoken by Lao-Tze.
If we are to believe that Lao-Tze
Was himself one who knew,
How comes it that he wrote a book
Of five thousand words?
Translated by Arthur Waley *Chinese Poems* 1946 (London: Allen & Unwin)

Edgar Allen Poe 1809–1849

77   Science! True daughter of Old Time thou art!
Who alterest all things with thy peering eyes...
Hast thou not dragged Diana from her car?
Sonnet *To Science* ca 1827

78   Let us examine a crystal...the equality of the sides pleases us; that of the angles doubles the pleasure. On bringing to view a second face in all respects similar to the first, this pleasure seems to be squared; and bringing into view a third, it appears to be cubed, and so on.
[$S = K \log W$ by intuition?]
The Rationale of Verse 1843

79   Scientific music has no claims to intrinsic excellence as it is fit for scientific ears only. In its excess it is the triumph of the physique over the morale of music—the sentiment is overwhelmed by the sense.
The Rationale of Verse 1843

Jules Henri Poincaré 1854–1912

80   *Les faits ne parlent pas.*
Facts do not speak.

81   Science is built up with facts, as a house is with stones. But a collection of facts is no more a science than a heap of stones is a house.
La Science at l'Hypothèse 1902 (Paris: Flammarion)

82   *La liberté est pour la Science ce que l'air est pour l'animal.*
Freedom is for science what the air is for an animal.
Dernières Pensées (Paris: Flammarion) Chapter 8

83   Mathematical discoveries, small or great...are never born of spontaneous generation. They always presuppose a soil seeded with preliminary knowledge and well prepared by labour, both conscious and subconscious.

84   Thought is only a flash between two long nights, but this flash is everything.

85   *L'esprit n'use de sa faculté creatrice que quand l'expérience lui en impose la necessité.*
The mind uses its faculty for creativity only when experience forces it to do so.
La Science et l'Hypothèse 1902 (Paris: Flammarion) Introduction

Michael Polanyi 1891–1976

86   The pursuit of science can be organized...in no other manner than by granting complete independence to all mature scientists. They will then distribute themselves over the whole field of possible discoveries, each applying his own special ability to the task that appears most profitable to him. The function of public authorities is not to plan research, but only to provide opportunities for its pursuit. All they have to do is to provide facilities for every good scientist to follow his own interest in science.
The Logic of Liberty 1951 (Chicago, IL: University of Chicago Press)

87   The Republic of Science shows us an association of independent initiatives, combined towards an indeterminate achievement. It is disciplined and motivated by serving a traditional authority, but this authority is dynamic; its continued existence depends on its constant self-renewal through the originality of its followers.
The Republic of Science is a Society of Explorers. Such a society strives towards an unknown future, which it believes to be accessible and worth achieving. In the case of scientists, the explorers strive towards a hidden reality, for the sake of intellectual satisfaction. And as they satisfy themselves,

they enlighten all men and are thus helping society to fulfil its obligation towards intellectual self-improvement.
Minerva 1962 1 54–73

Georges Politzer 1903–1942

88   *Les psychologues sont scientifiques comme les sauvages evangelisés sont chrêtiens.*
Psychologists are scientists about as much as converted savages are Christians.
Critique des Fondements de la Psychologie (Paris: P U F) Introduction

Marco Polo ca 1254–1324

89   Here is told of the city of Kinsai [Hangkow].... In other streets, live harlots, of whom there are so many that I dare not say the number.... These women are very clever and expert in their allurements and endearments, and ever have appropriate words ready for every kind of person. So, when foreigners have once tasted of them, they remain, so to speak, beside themselves, and are so taken by their sweetness and charm, that they can never forget them. Thus it is that, when they return home, they say they have been in Kinsai, namely in the City of Heaven, and long to be able to return there. In yet other streets live all the leeches and all the astrologers, the latter of whom also teach reading and writing.
The Travels of Marco Polo ed L F Benedetto, transl A Ricci, 1931 (London: Routledge & Kegan Paul)

Polybius ca 204–ca 122 BC

90   Whenever it is possible to find out the cause of what is happening, one should not have recourse to the gods.
In K Von Fritz *The Theory of the Mixed Constitution in Antiquity* 1954 (New York: Columbia University Press)

[Saint] Polycarp ca 09–ca 155

91   In all these monstrous demons is seen an art hostile to God.
[On the clepsydra]

Georges Pompidou 1911–1974

92   There are three roads to ruin; women, gambling and technicians. The most pleasant is with women, the quickest is with gambling, but the surest is with technicians.
Sunday Telegraph 26 May 1968

Alexander Pope 1688–1744

93   Epitaph on Newton:
Nature and Nature's laws lay hid in night:
God said, 'Let Newton be' and all was light.
[Added by Sir John Collings Squire:
It did not last: the Devil, shouting 'Ho.
Let Einstein be' restored the status quo.]
The Works of Alexander Pope 1871 (London: Murray)

94   For Forms of Government let fools contest;
What'er is best administered is best.
An Essay on Man III, line 303

95   How index-learning turns no student pale,
      Yet holds the eel of science by the tail.
      The Dunciad Book I, line 279

96   Lo, the poor Indian: whose untutored mind
      Sees God in clouds, or hears him in the wind:
      His soul proud science never taught to stray
      Far as the Solar Walk or Milky Way.
      An Essay on Man I, line 99

97   Not chaos-like together wash'd and bruis'd,
      But, as the world, harmoniously confus'd:
      Where order in variety we see,
      And where, though all things differ, all agree.
      Windsor Forest

98   One science only will one genius fit;
      So vast is art, so narrow human wit.
      An Essay on Criticism part I, line 60

99   Order is Heaven's first law.
      An Essay on Man IV, line 49

100   Those Rules of old discovered, not devised
      Are Nature still, but Nature methodized;
      Nature, like Liberty, is but restrained
      By the same Laws which first herself ordained.
      Perspectives in Biology and Medicine 1971, Autumn, p 105

101   Why has not man a microscopic eye?
      For this plain reason, man is not a fly.
      An Essay on Man I, 193

102   Atoms or systems into ruins hurl'd,
      And now a bubble burst, and now a World.
      An Essay on Man

[Sir] Karl Raimund Popper 1902–

103   But I shall certainly admit a system as empirical or scientific only if it is
      capable of being *tested* by experience. These considerations suggest that
      not the *verifiability* but the *falsifiability* of a system is to be taken as a
      criterion of demarcation. In other words: I shall not require of a scientific
      system that it shall be capable of being singled out, once and for all, in a
      positive sense: but I shall require that its logical form shall be such that it
      can be singled out, by means of empirical tests, in a negative sense: *it must
      be possible for an empirical scientific system to be refuted by experience.*
      The Logic of Scientific Discovery 1959 (London: Hutchinson)

104   Science is not a system of certain, or well-established, statements; nor is
      it a system which steadily advances towards a state of finality. ...And our
      guesses are guided by the unscientific the metaphysical (though biologically
      explicable) faith in laws, in regularities which we can uncover—discover.
      Like Bacon, we might describe our own contemporary science—'the method
      of reasoning which men now ordinarily apply to nature'—as consisting of
      'anticipations, rash and premature' and as 'prejudices'.
      The Logic of Scientific Discovery 1959 (London: Hutchinson)

105   What really makes science grow is new ideas, including false ideas.

106   According to positivism, the world is all surface.
      The Sciences 1979 **19**, No 10, 29

107   Big science may destroy great science.
      In R Harré (ed) *Problems of Scientific Revolution* 1975 (Oxford: Oxford University Press)

[Sir] George Porter 1920–

108   There is then, one great purpose for man and for us today, and that is to try to discover man's purpose by every means in our power. That is the ultimate relevance of science, and not only of science, but of every branch of learning which can improve our understanding. In the words of Tolstoy, 'The highest wisdom has but one science, the science of the whole, the science explaining the Creation and man's place in it'.
      The Times Saturday 21 June 1975

François Poullain de la Barre

109   The mind has no sex.
      De l'éducation des dames pour la conduite de l'esprit dans les sciences 1673

Ezra Pound 1885–1972

110   'You damn sadist,' said mr cummings,
      'you try to make people think.'
      Canto 89 in *The Cantos of Ezra Pound* 1956 (London: Faber & Faber and New York: New Directions) ©Ezra Pound, 1956

111   Of all those young women not one has enquired the cause of the world
      Nor the modus of lunar eclipses.
      Homage to Sextus Propertius in *The Collected Shorter Poems of Ezra Pound* 1926 (London: Faber & Faber and New York: New Directions) ©Ezra Pound, 1926

112   A Gnomon,
      Our science is from the watching of shadows...
      Canto 85 in *The Cantos of Ezra Pound* 1956 (London: Faber & Faber and New York: New Directions) ©Ezra Pound, 1956

113   The army vocabulary contained almost 48 words
      one verb and participle one substantive (hole—Gk)
      one adjective and one phrase sexless that is
      used as a sort of pronoun.
      Cantos 77 471:50 in *The Cantos of Ezra Pound* 1956 (London: Faber & Faber and New York: New Directions) ©Ezra Pound, 1926

114   [On the poetic image]
      A radiant node or cluster,...what I can, and must perforce, call a VORTEX, from which, and through which, ideas are constantly rushing.
      1914. In Hugh Kenner *The Pound Era* 1972 (London: Faber & Faber)

115   And what man of unusual intelligence in our day, or in any day, has been content to live away from, or out of touch with, the biggest metropolis he could get to?
      Selected Prose, 1909–1965 1973 (London: Faber & Faber)

116   Disciples are more trouble than they are worth when they start anchoring and petrifying their Mahatmas. No man's thought petrifies.
      Selected Prose, 1909–1965 1973 (London: Faber & Faber)

117   The toxicology of money.
      Selected Prose, 1909–1965 1973 (London: Faber & Faber)

Enoch Powell 1912–

118   In my branch of learning [history] experiment is impossible. Advance
      is made through thought. Experiment is often used as a substitute for
      thought.
      The Guardian 22 April 1985

Cedric Price 1934–

119   The reason for architecture is to encourage...people...to behave, mentally
      and physically, in ways they had previously thought impossible.
      Exhibition, RIBA Heinz Gallery, 8 October 1975

Derek John de Solla Price 1922–1983

120   The disciplines which analyse science have been generated piecemeal, but
      show many signs of beginning to cohere into a whole which is greater than
      the sum of its parts. This new study might be called 'history, philosophy,
      sociology, psychology, economics, political science and operations research
      (etc) of science, technology, medicine (etc).
      In *The Science of Science* ed M Goldsmith and A L Mackay, 1964 (London: Souvenir)

121   Science is not just the fruit of the tree of knowledge, it is the tree itself.
      Lecture, London, 1964

122   Using any reasonable definition of a scientist, we can say that between 80
      and 90 per cent of all the scientists that have ever lived are alive now.
      Now depending on what one measures and how, the crude size of science

in manpower or in publications tends to double within a period of 10 to 15 years.
Little Science, Big Science 1963 (New York: Columbia University Press)

Don Krasher Price 1910–

123   Science...cannot exist on the basis of a treaty of strict non-aggression with the rest of society; from either side, there is no defensible frontier.
Government and Science 1954 (New York: New York University Press)

Joseph Priestley 1733–1804

124   It was ill policy in Leo the Tenth to patronise polite literature. He was cherishing an enemy in disguise. And the English hierarchy (if there be anything unsound in its constitution) has equal reason to tremble even at an air pump or an electrical machine.
Experiments and Observations on Different Kinds of Air 1775–1786

Proclus Diadochus 412–485

125   It is well known that the man who first made public the theory of irrationals perished in a shipwreck in order that the inexpressible and unimaginable should ever remain veiled. And so the guilty man, who fortuitously touched on and revealed this aspect of living things, was taken to the place where he began and there is for ever beaten by the waves.
[Attributed]
Scholium to Book X of *Euclid* V, 417

Protagoras ca 481–ca 411 BC

126   Man is the measure of all things, of things that are, that they are, of things that are not, that they are not.
In Diogenes Laertius *Vitae Philosophicus* IX, 51

127   Of the gods I know nothing, whether they exist or do not exist: nor what they are like in form. Many things stand in the way of knowledge—the obscurity of the subject, the brevity of human life.
In Diogenes Laertius *Vitae Philosophicus* IX, 51

Pierre Joseph Proudhon 1809–1865

128   *[De même qu'il y a] une science des phenomènes physiques qui ne repose que sur l'observation des faits, il doit exister aussi une science de la société, absolue, rigoureuse, basée sur la nature de l'homme et de ses facultés, et sur leurs rapports, science qu'il ne faut pas inventer mais découvrir.*
Inasmuch as there is a science of physical phenomena, which rests only on the observation of facts, there ought also to exist a science of society which should be absolute and rigorous and based on the nature of man, his faculties and their inter-relationships. This should be a science to be discovered, not invented.
L'Utilité de la Célébration du Dimanche 1839

Marcel Proust 1871–1922

129   Distances are only the relation of space to time and vary with that relation.
Cities of the Plain, I, 3

130   Theories and schools, like microbes and globules, devour each other and by their struggle ensure the continuing of life.
Sodome et Gomorre

Aleksandr Sergeyevich Pushkin 1799–1837

131   What is now the custom in London is still premature in Moscow.
First Submission to the Censor 1822. In *Collected Works*

132   For how many marvellous discoveries, do the enlightenment of souls, and experiment, son of serious errors, and genius, the friend of paradox and chance, the god-inventor, prepare us.
Likhtenshtein 209

133   Science shortens for us the experimentation needed in a fast-flowing life.

134   Inspiration is needed in geometry, just as much as in poetry.
Likhtenshtein 266

Pythagoras 6th Century BC

135   Number is the ruler of forms and ideas, and the cause of gods and demons.
From *Iamblichus* by R Graves *The White Goddess* 3rd edn, 1952

François Quesnay 1694–1774

1   Commerce, like industry, is merely a branch of agriculture. It is agriculture
which furnishes the material of industry and commerce and which pays
both...
Grains in *Encyclopédie*

Adolphe Quetelet 1796–1874

2   'The average man'.
[The invention of the concept]

3   The more progress physical sciences make, the more they tend to enter
the domain of mathematics, which is a kind of centre to which they all
converge. We may even judge of the degree of perfection to which a science
has arrived by the facility with which it may be submitted to calculation.
In E Mailly *Eulogy on Quetelet* 1874, Smithsonian Report

John Quincy

4   CRYSTALLIZATION: such a combination of saline particles as resembles
the form of a crystal, variously modified, according to the nature and texture of the salts. The method is by dissolving any saline body in water,
and filtering it, to evaporate, till a film appears at the top, and then let
it stand to shoot; and this it does by that attractive force which is in all
bodies, and particularly in salt, by reason of its solidity: whereby, when
the menstruum or fluid, in which such particles flow, is sated enough or
evaporated, so that the saline particles are within each other's attractive
powers, they draw one another more than they are drawn by the fluid, then
will they run into crystals. And this is peculiar to those, that let them be
ever so much divided and reduced into minute particles, yet when they are
formed into crystals, they each of them reassume their proper shapes; so
that one might as easily divest them of their saltiness, as of their figure.
This being an immutable and perpetual law, by knowing the figure of the
crystals, we may understand what the texture of the particles ought to be,
which can form those crystals; and, on the other hand, by knowing the
texture of the particles, may be determined the figure of the crystals.
Probably *Lexicon Physico-Medicum* 1719; in *Samuel Johnson's Dictionary* 1755.
CRYSTALLIZATION

François Rabelais 1494–1553

1   *...questio subtillissima, utrum chimera in vacuo bombinans possit comedere secundas intentiones.*
...a most subtle question whether a chimaera bombinating in a vacuum can devour second intentions. (Oxford English Dictionary.)
Tiers Livre Book II, vii

2   Nature abhors a vacuum.
[Quoting the Latin proverb *natura vacuum abhorret*]
Gargantua Book 1, Chapter 5

3   Science without conscience is but the ruin of the soul.
Gargantua's letter to Pantagruel

Isador Isaac Rabi 1898–1987

4   There isn't a scientific community. It is a culture. It is a very undisciplined organisation.
1965. In D S Greenberg *The Politics of Pure Science* 1967 (New York: New American Library)

[Sir] Walter Alexander Raleigh 1861–1922

5   In an examination those who do not wish to know ask questions of those who cannot tell.
Some Thoughts on Examinations

Srinivasa Ramanujan 1887–1920

6   G H Hardy: I remember once going to see him when he was lying ill at Putney. I had ridden in taxi-cab No.1729 and remarked that the number seemed to me rather a dull one, and that I hoped it was not an unfavourable omen. 'No,' he replied, 'it is a very interesting number; it is the smallest number expressible as the sum of two cubes in two different ways.'
In G H Hardy *Ramanujan* 1940 (London: Cambridge University Press)

James Arthur Ramsay 1909–1987

7   The mammal is a highly-tuned physiological machine carrying out with superlative efficiency what the lower animals are content to muddle through with.

[Sir] William Mitchell Ramsay 1851–1939

8   This war, in contradistinction to all previous wars, is a war in which pure and applied science plays a conspicuous part.
     Nature 1915 **95** 521

Anatol Rapoport 1911–

9   One cannot play chess if one becomes aware of the pieces as living souls and of the fact that the Whites and the Blacks have more in common with each other than with the players. Suddenly one loses all interest in who will be champion.
     Strategy and Conscience 1964 (New York: Harper & Row)

Dixy Lee Ray 1914–

10   [U S Minister] The general public has long been divided into two parts; those who think science can do anything, and those who are afraid it will.
     New Scientist 5 July 1973 **59** (853) 14

John William Strutt [Lord] Rayleigh 1842–1919

11   Some proofs command assent. Others woo and charm the intellect. They evoke delight and an overpowering desire to say 'Amen, Amen'.
     In H E Hunter *The Divine Proportion* 1970 (New York: Dover)

Herbert Read 1893–1968

12   Beauty had been born, not, as we so often conceive it nowadays, as an ideal of humanity, but a **measure**, as the reduction of the chaos of appearances to the precision of linear symbols. Symmetry, balance, harmonic division, mated and measured intervals— such were its abstract characteristics.
     Icon and Idea 1955 (London: Faber & Faber)

Ronald Reagan 1911–

13   [Then (1981–1989) President of the USA—On evolution] Well, it is a theory, it is a scientific theory only, and it has in recent years been challenged in the world of science and is not yet believed in the scientific community to be as infallible as it was once believed. But if it was going to be taught in the schools, then I think that also the biblical theory of creation, which is not a theory but the biblical theory of creation, should also be taught.
     In M Rusé *Is Science Sexist?* 1981 (Dordrecht: Reidel)

14   Now, therefore, I Ronald Reagan, President of the United States of America, do hereby proclaim the week of April 14 through April 20 1986, as National Mathematics Awareness Week, and I urge all Americans to participate in appropriate ceremonies and activities that demonstrate the importance of mathematics and mathematical education to the United States.
     Proclamation 5461 of 17 April 1986 [thus 3 days late!]

Hans Reichenbach 1891–1953

15   If error is corrected whenever it is recognised as such, the path of error is the path of truth.

Alastair Reid 1926–

16   'Counting': Ounce, dice, trice, quartz, quince, sago, serpent, oxygen, nitrogen, denim.
     Ounce, dice, trice 1956 (Boston, MA: Little, Brown)

Georg Joachim Rheticus 1514–1576

17   The planets show again and again all the phenomena which God desired
to be seen from the Earth.
In O Neugebauer *The Exact Sciences in Antiquity* 1957

Ivor Armstrong Richards 1893–1979

18   The properties of the instruments or apparatus employed enter
into...belong with and confine the scope of the investigation.
Speculative Instruments 114 in *Internal Colloquies* (London: Routledge & Kegan
Paul)

19   The despondencies, the emotional excitements accompanying research and
discovery [are] again on an unprecedented scale. Thus, a number of men who
might in other times have been poets are today in bio-chemical laboratories.
Science and Poetry 1926 (London: Kegan Paul)

Lewis Fry Richardson 1881–1953

20   Big[ger] whirls have little whirls,
That feed on their velocity;
And little whirls have lesser whirls,
And so on to viscosity.
[Summarizing his classic paper *The Supply of Energy from and to Atmospheric Eddies*
(1920)]

Charles Robert Richet 1850–1935

21   I possess every good quality, but the one that distinguishes me above all is
modesty.
[Nobel Laureat for medicine, 1913]
The Natural History of a Savant transl Oliver Lodge, 1927 (New York: Doran)

F K Richtmeyer

22   The whole history of physics proves that a new discovery is quite likely
lurking at the next decimal place.
Science 1932 **75** 1–5

Georg Friedrich Bernhard Riemann 1826–1866

23   Therefore, either the reality on which our space is based must form a dis-
crete manifold or else the reason for the metric relationships must be sought
for, externally, in the binding forces acting upon it.
Lecture on the foundations of geometry. In S Chandrasekhar *Nature* 22 March 1990
344 285

Arthur Rimbaud 1854–1891

24   *Science avec patience: Le supplice est sur.*
Science with patience. Torture is certain.
L'Éternité 1872

[Bishop of] Ripon

25   Let us halt scientific research for ten years.
In B M Caroe *William Henry Bragg* 1978 (Cambridge: Cambridge University Press)

[Baron] Peter Ritchie-Calder 1906–1982

26   The academic and basic scientists are 'The Makers-Possible', the applied
scientists and the technologists are 'The Makers-to-Happen', and the technicians 'The Makers-to-Work'. And nowadays, with operations research,
market research, quality control, etc, the commercial scientists might be
called 'The Makers-to-Pay'.
The Evolution of Science 1963 (Paris: UNESCO/Mentor)

Michael Roberts 1902–1948

27   ...seize
The elusive photon's properties
In α's and δ's, set in bronze—
Bright vectors, grim quaternions.
Note on θ, ϕ and ψ

José Roderigues Migueis

28   Creativity is what cannot wait, cannot stop, cannot backstep: faster or
slower, it always goes ahead—through, alongside, above, regardless of crises
or systems.
Address to the Center for Portuguese and Brazilian Studies, Brown University, 11
March 1979. In June Goodfield *An Imagined World*

Jacques Rohault 17th Century

29   ...it was by just such a hazard, as if a man should let fall a handful of sand
upon a table and the particles of it should be so ranged that we could read
distinctly on it a whole page of Virgil's *Aenead*.
Traité de Physique Paris, 1671. Transl 1723, II

Jules Romains 1885–

30   *Les gens bien portants sont des malades qui s'ignorent.*
Every man who feels well is a sick man neglecting himself.
Knock, ou le Triomphe de la Médecine 1923 (Paris: Gallimard)

A Rosenblueth

31   [With Norbert Wiener]
The best material model of a cat is another, or preferably the same, cat.
Philosophy of Science 1945 **12**

Dante Gabriel Rosetti 1828 1882

32   Conception, my boy, fundamental brainwork, is what makes the difference
in all art.
Letter to Hall Caine

[Sir] Ronald Ross 1857–1932

33   This day relenting God
Hath placed within my hand
A wondrous thing; and God
Be praised. At His command,
Seeking His secret deeds
With tears and toiling breath,
I find thy cunning seeds,

O million-murdering Death.
I know this little thing
A myriad men will save
O Death where is thy sting?
Thy victory, O Grave?
[Describing his discovery of the life-cycle of the malaria parasite. 1897]
Poems by Ronald Ross 1928 (London: E Mathews & Marrot)

34   Now twenty years ago
This day we found the thing;
With science and with skill
We found; then came the sting—
What we with endless labour won
The thick world scorned;
Not worth a word today—
Not worth remembering.
[Ross received the Nobel Prize for medicine in 1902]
Written 20 August 1917

Jean Rostand 1894–

35   *Ou apprendre le métier de Dieu?*
Who can teach us God's business?

36   *Être adulte, c'est être seul.*
To be adult is to be alone.
Pensées d'un biologiste (Paris: Stock)

37   *Récherche scientifique; la seule forme de poésie qui soit retribuée par l'état.*
Scientific research; the only form of poetry rewarded by the state.
Inquiètudes d'un biologiste, 2, Stock

38   *Le masculin est mêlé de fémininité, le feminin est pur.*
The masculine has an admixture of femininity: the feminine is pure.
Inquiètudes d'un Biologiste, 2, Stock

Theodore Roszak 1933–

39   Nature composes some of her loveliest poems for the microscope and telescope.
Where the Wasteland Ends 1972 (London: Faber & Faber)

Jean-Jacques Rousseau 1712–1778

40   [Rousseau curiously enough, argued that the size of the elite (i.e. government) varied with the square root of the population.]
1762. *Social Contract* (Oxford: Oxford)

Joseph Roux 1834–1886

41   Science is for those who learn; poetry for those who know.
Meditations of a Parish Priest 1

Henry Augustus Rowland 1848–1901

42   He who makes two blades of grass grow where one grew before is the benefactor of mankind, but he who obscurely worked to find the laws of such growth is the intellectual superior as well as the greater benefactor of mankind.
In D S Greenberg *The Politics of Pure Science* 1967 (New York: New American Library)

Josiah Royce 1855–1916

43 Let us imagine that a portion of the soil of England has been levelled off perfectly and that on it a cartographer traces a map of England. The job is perfect; there is no detail of the soil of England, no matter how minute, that is not registered on the map; everything has there its correspondence. This map, in such a case, should contain a map of the map, which should contain a map of the map, and so on to infinity.
[See the model of the village of Bourton-on-the-Water in Bourton-on-the-Water...]
The World and the Individual (Gifford Lectures) 1900–1901

Count Rumford [Benjamin Thompson] 1753–1814

44 It frequently happens that in the ordinary affairs and occupations of life, opportunities present themselves of contemplating some of the most curious operations of nature.
An Inquiry Concerning the Source of the Heat which is Excited by Friction. Read to the Royal Society on 25 January 1978

Benjamin Rush 18th Century

45 Man is said to be a compound of soul and body. However proper this language may be in religion, it is not so in medicine. He is, in the eye of a physician, a single and indivisible being, for so intimately united are his soul and body, that one cannot be moved without the other.
Benjamin Rush *Sixteen introductory lectures* 1811 (Philadelphia, Pa: Bradford and Innskeep)

Salman Rushdie 1947–

46 By the time the rain came at the end of June, the foetus was fully formed inside her womb. Knees and nose were present; and as many heads as would grow were already in position. What had been (at the beginning) no bigger than a full stop had expanded into a comma, a word, a sentence, a paragraph, a chapter; now it was bursting into more complex developments, becoming, one might say, a book—perhaps an encyclopaedia—even a whole language...
Midnight's Children 1982 (London: Picador)

Dean Rusk 1909–

47 [US Secretary of State 1961–]
The world is round. Only one third of its people are asleep at any one time. The other two thirds are awake and causing mischief somewhere.

John Ruskin 1819–1900

48 *Lily:* 'We looked at the books about crystals but they are so dreadful.'
The Ethics of the Dust, Ten Lectures to Little Housewives on the Elements of Crystallisation 1866 (London: Smith Elder)

49 *May:* 'Oh. Have the crystals faults like us?'
L: 'Certainly, May. Their best virtues are shown in fighting their faults. And some have a great many faults; and some are very naughty crystals indeed.'
The Ethics of the Dust, Ten Lectures to Little Housewives on the Elements of Crystallisation 1866 (London: Smith Elder)

50   [The crystal's] limited, though...stern, code of morals.
      The Ethics of the Dust. Ten Lectures to Little Housewives on the Elements of Crystallisation 1866 (London: Smith Elder)

51   The use of the word **scientia**, as if it differed from **knowledge**, [is] a modern barbarism; enhanced usually by the assumption that the knowledge of the difference between acids and alkalis is a more respectable one than that of the difference between vice and virtue.
      [Ruskin resigned his professorial chair at Oxford in 1884 as an expression of disapproval when a laboratory for physiology was established in the university.]
      Works (ed) 1908 (London: Cook and Wedderburn) vol 34

52   Life without industry is guilt, and industry without art is brutality.
      Lectures on Art

[Lord] Bertrand Russell 1872–1970

53   ...the general public has derived the impression that physics confirms practically the whole of the Book of Genesis. I do not myself think that the moral to be drawn from modern science is at all what the general public has thus been led to suppose. In the first place, the men of science have not said nearly as much as they are thought to have said, and in the second place what they have said in the way of support for traditional religious beliefs has been said by them not in their cautious, scientific capacity, but rather in their capacity of good citizens, anxious to defend virtue and property.
      The Scientific Outlook 1931 (London: Allen & Unwin)

54   Animals studied by Americans rush about frantically, with an incredible display of hustle and pep, and at last achieve the desired result by chance. Animals observed by Germans sit still and think, and at last evolve the solution out of their inner consciousness.
      [On Thorndike and Koehler]

55   Aristotle maintained that women have fewer teeth than men; although he was twice married, it never occurred to him to verify this statement by examining his wives' mouths.
      The Impact of Science on Society 1952 (London: Allen & Unwin)

56   Can a society in which thought and technique are scientific persist for a long period, as, for example, ancient Egypt persisted, or does it necessarily contain within itself forces which must bring either decay or explosion...?
      Lloyd Roberts Lecture *Can a Scientific Community be Stable?* to the Royal Society of Medicine on 29 November 1949

57   I am compelled to fear that science will be used to promote the power of dominant groups rather than to make men happy.
      [Replying to J B S Haldane's optimistic view expressed in *Daedalus—Science and the Future*]
      Icarus, or the Future of Science 1925 (London: Kegan Paul)

58   The number of a class is the class of all classes similar to a given class.
      Principles of Mathematics 1903 (New York: Cambridge University Press)

59   Pure mathematics is the class of all propositions of the form p implies q, where p and q are propositions containing one or more variables, the same in the two propositions, and neither p nor q contains any constants except logical constants.... The fact that all Mathematics is Symbolic Logic is

one of the greatest discoveries of our age; and when this fact has been established, the remainder of the principles of mathematics consists in the analysis of Symbolic Logic itself.
Principles of Mathematics 1903 (New York: Cambridge University Press)

60 The fundamental concept in social science is Power, in the same sense in which Energy is the fundamental concept in physics.
Power: A New Social Analysis 1938 (New York: Norton)

61 With equal passion I have sought knowledge. I have wished to understand the hearts of men. I have wished to know why the stars shine. And I have tried to apprehend the Pythagorean power by which number holds sway above the flux. A little of this, but not much, I have achieved.
The Autobiography of Bertrand Russell (London: Allen & Unwin). Introduction

62 Ordinary language is totally unsuited for expressing what physics really asserts, since the words of everyday life are not sufficiently abstract. Only mathematics and mathematical logic can say as little as the physicist means to say.
The Scientific Outlook 1931 (London: Allen & Unwin)

63 A good notation has a subtlety and suggestiveness which at times make it almost seem like a live teacher.

64 At the age of eleven, I began Euclid, with my brother as my tutor. This was one of the great events of my life, as dazzling as first love. I had not imagined there was anything so delicious in the world.... From that moment until...I was thirty-eight, mathematics was my chief interest and my chief source of happiness.
Autobiography

65 Mathematics may be defined as the subject in which we never know what we are talking about, nor whether what we are saying is true.
Mathematics and the Metaphysicians in *Mysticism and Logic* 1917 Chapter 4

66 The man who has fed the chicken every day throughout its life at last wrings its neck instead, showing that more refined views as to the uniformity of nature would have been useful to the chicken.
The Problems of Philosophy 1936 (London: Thornton, Butterworth)

67 Even if the open windows of science at first make us shiver after the cosy indoor warmth of traditional humanizing myths, in the end the fresh air brings vigour, and the great spaces have a splendour of their own.
What I believe 1925

Henry Norris Russell 1877–1957

68 The pursuit of an idea is as exciting as the pursuit of a whale.

[Lord] Ernest Rutherford 1871–1937

69 Don't let me catch anyone talking about the Universe in my department.
John Kendrew, BBC-3, 26 July 1968, 21.00 h
J D Bernal and the Origin of Life

70 If your experiment needs statistics, you ought to have done a better experiment.
In N T J Bailey *The Mathematical Approach to Biology and Medicine* 1967 (New York: Wiley)

71   It is essential for men of science
to take an interest in the
administration of their own affairs
or else the professional civil servant
will step in—and then the Lord
help you.
Bulletin of the Institute of Physics 1950
1 No 1, cover

72   We haven't the money, so we've got
to think.
In R V Jones *Bulletin of the Institute
of Physics* 1962 **13** No 4 102
B V Bowden, *New Scientist* 30
September 1965

73   [In answer to Stephen Leacock's enquiry as to what he thought of Einstein's
theory of relativity] Oh, that stuff. We never bother with that in our work.
In Stephen Leacock *Common Sense and the Universe*

74   The energy produced by the breaking down of the atom is a very poor kind
of thing. Anyone who expects a source of power from the transformation
of these atoms is talking moonshine.
Physics Today 1970, October, p 33; from *New York Herald Tribune* 12 September
1933

75   Science is divided into two categories, physics and stamp-collecting.
In J D Bernal *The Social Function of Science* 1939

76   Never say, 'I tried it once and it did not work'.

Gilbert Ryle 1900–1976

77   [There is no] Ghost in the Machine.
The Concept of Mind 1949 (London: Hutchinson)

George Alban Sacher 1917–1981

1   The brain is the organ of longevity.
[By its capacity to regulate the *milieu intérieur*]
Perspectives in Experimental Gerontology 1966 (Springfield, IL: Thomas)

Donatin Alphonse François [Marquis de] Sade 1740–1814

2   *A quelque point qu'en frémissant les hommes, la philosophie doit tout dire.*
No matter how much this may shock mankind, the duty of philosophy is
to say everything.
Juliette, ou la prosperité du vice 1797

Carl Sagan 1934–

3   Our loyalties are to the species and the planet. *We* speak for Earth. Our
obligation to survive is owed not just to ourselves but also to that Cosmos,
ancient and vast, from which we spring.
Cosmos 1980 (London: Macdonald)

4   Almost half the scientists and high technologists on Earth are employed full-
or part-time on military matters. Those engaged in the development and
manufacture of weapons of mass destruction are given salaries, perquisites
of power and, where possible, public honors at the highest levels in their
respective societies. The secrecy of weapons development, carried to especially extravagant lengths in the Soviet Union, implies that individuals so
employed almost never need accept responsibility for their actions. They
are protected and anonymous. Military secrecy makes the military the most
difficult sector of any society for the citizens to monitor. If we do not know
what they do, it is very hard for us to stop them. And with the rewards
so substantial, with the hostile military establishments beholden to each
other in some ghastly embrace, the world discovers itself drifting toward
the ultimate undoing of the human enterprise.
Cosmos 1080 (London: Macdonald)

Antoine de Saint-Exupéry 1900–1944

5   *L'avion est une machine sans doute, mais quel instrument d'analyse! Cet
instrument nous a fait découvrir le vrai visage de la terre.... Nous voilà
donc changés en physiciens, en biologistes, examinant ces civilisations qui
ornent des fonds de vallées.... Nous voilà donc jugeant l'homme a l'échelle
cosmique, l'observant a travers nos hublots, comme à travers des instruments d'étude. Nous voilà relisant notre histoire.*
The aeroplane is, of course, only a machine, but what an instrument of
analysis! This instrument has made us see the real face of the earth....Up
here we are turned into physicists or biologists, studying the civilizations
which garnish the depths of the valleys....Up here we are judging man on
a cosmic scale, observing him through our portholes, as through scientific
instruments. We are re-reading our own history.
Oeuvres d'Antoine de Saint-Exupéry 1953 (Paris: Gallimard)

6   When the body sinks into death, the essence of man is revealed. Man is a
knot, a web, a mesh into which relationships are tied. Only those relationships matter. The body is an old crock that nobody will miss. I have never
known a man to think of himself when dying. Never.
Flight to Arras transl Lewis Galantiere (London: Heinemann)

Andrei Dmitrievich Sakharov 1921–1989

7   Every day I saw the huge material, intellectual and nervous resources of thousands of people being poured into the creation of a means of total destruction, something capable of annihilating all human civilisation. I noticed that the control levers were in the hands of people who, though talented in their own ways, were cynical. Until the summer of 1953 the chief of the atomic project was Beria, who ruled over millions of slave-prisoners. Almost all the construction was done with their labour. Beginning in the late fifties, one got an increasingly clearer picture of the collective might of the military–industrial complex and of its vigorous, unprincipled leaders, blind to everything except their 'job'.
[Later, as in the case of Galileo, they showed him the instruments of torture (for forcible feeding) and later subjected him to this process]
Sakharov Speaks 1974 (London: Harvill Press)

Sakuma Shozan 1811–1864

8   *Toyo dotoku, seiyo geijutsu.*
Eastern ethics, Western techniques.
[The conditions for modernization at the Meiji Restoration]
[See Chang Chih-tung]

Abdus Salam 1926–

9   Al Asuli writing in Bukhara some 900 years ago divided his pharmacopoeia into two parts, 'Diseases of the rich' and 'Diseases of the poor'.
Scientific World 1963, No 3, p9

Denis de Sallo 1626–1669

10   [First editor of the first scientific journal, writing in the first issue]
Personne ne doit trouver estrange de voir ici des opinions différentes des siennes, touchant les sciences, puisqu'on fait profession de rapporter les sentiments des autres sans les garantir....
Nobody should find it strange to see here opinions different from his own concerning the sciences, because we aim to report the ideas of others without guaranteeing them....
Journal des Scavans 1665

Saif-ud-din Salman 15th Century

11   [Working at the observatory of Ulugh Beg in Samarkand] Admonish me not, my beloved father, for forsaking you thus in your old age and sojourning here at Samarkand. It is not that I covet the musk melons and the grapes and the pomegranates of Samarkand; it is not the shades of orchards on the banks of Zar-Afsham, that keep me here. I love my native Kandahar and its tree-lined avenues even more and I pine to return. But forgive me, my exalted father, for my passion for knowledge. In Kandahar there are no scholars, no libraries, no quadrants, no astrolabes. My stargazing excites nothing but ridicule and scorn. My countrymen care more for the glitter of the sword than for the quill of the scholar. In my own town I am a sad, a pathetic misfit. It is true, my respected father, so far from home men do not rise from their seats to pay me homage when I ride into the bazaar.

But some day soon, all Samarkand will rise in respect when your son will emulate Biruni and Tusi in learning and you too will feel proud.
[Abdus Salam tells us that, alas! Salman never did get his PhD]
Transl Abdus Salam in *Minerva* 1966 4 461–5

Carl Sandburg 1878–1967

12   When water turns ice does it remember one time it was water? When ice turns back into water does it remember it was ice?
Metamorphosis in *Honey and Salt* 1963 (New York: Harcourt, Brace & World)
©Carl Sandburg, 1963

George Santayana 1863–1952

13   The empiricist...thinks he believes only what he sees, but he is much better at believing than at seeing.
Skepticism and Animal Faith 1955 (New York: Dover)

14   Those who cannot remember the past are condemned to repeat it.
The Life of Reason 1905 (New York: Scribners)

15   Great is this organism of mud and fire, terrible this vast, painful, glorious experiment.

George Alfred Leon Sarton 1884–1956

16   *Definition*: Science is systematised positive knowledge, or what has been taken as such at different ages and in different places.
Theorem: The acquisition and systematisation of positive knowledge are the only human activities which are truly cumulative and progressive.
Corollary: The history of science is the only history which can illustrate the progress of mankind. In fact, progress has no definite and unquestionable meaning in other fields than the field of science.
The Study of the History of Science 1957 (New York: Dover)

17   It is true that most men of letters and, I am sorry to add, not a few scientists, know science only by its material achievements, but ignore its spirit and see neither its internal beauty nor the beauty it extracts from the bosom of nature. Now I would say that to find in the works of science of the past, that which is not and cannot be superseded, is perhaps the most important part of our quest. A true humanist must know the life of science as he knows the life of art and the life of religion.
A History of Science vol II, 1959 (New York: Wiley)

18   The great intellectual division of mankind is not along geographical or racial lines, but between those who understand and practice the experimental method and those who do not understand and do not practice it.
In B Farrington *Science and Politics in the Ancient World* 1965 (London: Allen & Unwin)

19   Science always revolutionary and heterodox; it is its very essence to be so; it ceases to be so only when it is asleep.
In S F Mason *A History of the Sciences* 1953 (London: Routledge)

George Savile [Lord Halifax] 1633–1695

20   He that leaveth nothing to chance will do few things ill, but he will do very few things.
Complete Works of George Savile ed W Raleigh, 1912

21   The struggle for knowledge hath a pleasure in it like that of wrestling with a fine woman.
Complete Works of George Savile ed W Raleigh, 1912

Theodore H Savory 1896–

22   The action of a beakerful of hot nitric acid on half a pound of sugar is one of the fantastic sights in elementary chemistry and to watch it is an astonishing experience: I have known it to interrupt a nearby game of lawn-tennis.
The Language of Science 1953 (London: Deutsch)

A L Schawlow 1921–

23   To do successful research you don't need to know everything. You just need to know of the one thing that isn't known.
Springer house magazine

Friedrich von Schiller 1759–1805

24   *Einem ist sie [Wissenschaft] die hohe, die himmlische Göttin; dem anderen,*
Eine tüchtige Kuh, die ihn mit Butter versorgt.
To one science is an exalted goddess; to another it is a cow which provides him with butter.
Xenien

25   *Nur die Fülle führt zur Klarheit,*
Und im Abgrund wohnt die Wahrheit.
Only wholeness leads to clarity,
And truth lies in the abyss.
[Favourite saying of Niels Bohr]

26   The Greeks looked on Nature with their minds more than their hearts.

Arthur Schopenhauer 1788–1860

27   The chief objection I have to Pantheism is that it says nothing. To call the world 'God' is not to explain it: it is only to enrich our language with a superfluous synonym for the word 'world'.
A Few Words on Pantheism 1851

Erwin Schrödinger 1887–1961

28   ...a living organism...feeds upon negative entropy.... Thus the device by which an organism maintains itself stationary at a fairly high level of orderliness (=fairly low level of entropy) really consists in continually sucking orderliness from its environment.
What is Life? 1944 (London: Cambridge University Press)

29   Thus, the task is, not so much to see what no one has yet seen; but to think what nobody has yet thought, about that which everybody sees.
In L Bertalanffy *Problems of Life* 1952 (London: Watts)

30   Why must our bodies be so large compared with the atom?
[Loschmidt consisted of some 10^{27} atoms]
What is Life? 1944 (Cambridge: Cambridge University Press)

Russell Schweikart 1935–

31   The most beautiful sight in orbit, or one of the most beautiful sights, is a urine dump at sunset, because as the stuff comes out and as it hits the exit nozzle it instantly flashes into ten million little ice crystals which go out almost in a hemisphere, because, you know, you're exiting into essentially a perfect vacuum, and so the stuff goes in every direction, and all radially out from the space-craft at relatively high velocity. It's surprising, and it's an incredible stream of...just a spray of sparklers almost. It's really a spectacular sight. At any rate that's the urine system on Apollo.
The Next Whole Earth Catalogue 1981

Scottish Proverb

32   *Guid gear gangs intae sma bouk.*
Good equipment goes into small bulk.

Seki Takakazu [Kowa] 1642–1708

33   [Inventor of determinants]
Mathematics is more than an art form.
In S Nakayama *Dictionary of Scientific Biography* from Hirayama Akira, *Seki Takakazu*, 1959 (Tokyo: Koseisha)

Lucius Annaeus Seneca 4 BC–65 AD

34   In my own time there have been inventions of this sort, transparent windows, tubes for diffusing warmth equally through all parts of a building, short-hand, which has been carried to such a perfection that a writer can keep pace with the most rapid speaker. But the inventing of such things is drudgery for the lowest slaves; philosophy lies deeper. It is not her office to teach men how to use their hands. The object of her lessons is to form the soul. *Non est, inquam, instrumentorum ad usus necessarios opifex.*
[She is not the maker of tools for practical ends.]
Epistolae Morales 90

35   [Vice] is still in its infancy, and yet on it we bestow all our efforts: our eyes and our hands are its slaves. Who attends the school of wisdom now?...Who has regard for philosophy or any liberal pursuit, except when a rainy day comes round to interrupt the games, and it may be wasted without loss? And so the many sects of philosophers are dying out for lack of successors. The Academy, both old and new, has left no disciple.
In John Clarke *Physical Science* in *The Times of Nero: Being a Translation of the Quaestiones Naturales of Seneca* 1910 (London: Macmillan)

Severinus 7th Century

36   Go, my sons, buy stout shoes, climb the mountains, search...the deep recesses of the earth.... In this way and in no other will you arrive at a knowledge of the nature and properties of things.

William Shakespeare 1564–1616

37   Plutus himself,
      That knows the tinct and multiplying med'cine,
      Hath not in nature's mystery more science
      Than I have in this ring.
      All's Well That Ends Well V, iii

38   It is as easy to count atomies, as to resolve the propositions of a lover.
      As You LIke It III, ii

39   ...a traitorous innovator,
      A foe to the public weal.
      Coriolanus III, i

40   What doth gravity out of his bed at midnight?
      1 Henry IV II, iv

41   And time that takes survey of all the world
      Must have a stop.
      1 Henry IV V, iv

42   When we mean to build,
      We first survey the plot, then draw the model;
      And when we see the figure of the house,
      Then we must rate the cost of the erection;...
      2 Henry IV I, iii

43   Even so our houses and ourselves and children
      Have lost, or do not learn from want of time,
      The sciences that should become our country;
      But grow like savages,...
      Henry V V, ii

44   Thou hast most traitrously corrupted the youth
      Of the realm in erecting a grammar-school.
      2 Henry VI IV, vii

45   And nature must obey necessity.
      Julius Caesar IV, iii
      [See Democritus]

46   If you can look into the seeds of time,
      And say which grain will grow and which will not,
      Speak then to me...
      Macbeth I, iii

47   ...we but teach
      Bloody instructions, which being taught, return
      To plague the inventor.
      Macbeth I, vii

48   *Macbeth:* [The labour we delight] in physics pain.
      Macbeth II, iii

49   For there was never yet philosopher
      That could endure the toothache patiently.
      Much Ado about Nothing V, i

50   ...physics the subject, makes old hearts fresh;...
     The Winter's Tale I, i

51   *Sir Toby:* Does not our lives consist of the four elements?
     Sir Andrew: Faith, so they say; but I think it rather consists of eating
     and drinking.
     Sir Toby: Th'art a scholar; let us therefore eat and drink.
     Twelfth Night II, iii

52   In Nature's infinite book of secrecy
     A little can I read.
     Anthony and Cleopatra I.ii, 10

53   A good sherris-sack hath a two-fold operation in it... vapours...full of
     nimble fiery and delectable shapes.
     2 Henry IV IV, iii

54   The earth hath bubbles, as the water has,
     And these are of them.
     Macbeth I, iii

55   I cannot do it without comp[u]ters.
     The Winter's Tale IV, iii

56   I could be bounded in a nutshell
     And count myself a king of infinite space
     Were it not that I have bad dreams.
     Hamlet II, ii

57   There's no art
     To find the mind's construction in the face;
     Macbeth I, iv

58   Now the melancholy god protect thee; and the tailor make thy doublet of
     changeable taffeta, for thy mind is a very opal.
     Twelfth Night II, iv

59   Say, from whence
     You owe this strange intelligence?
     Macbeth I, iii

60   The king hath note of all that they intend,
     By interception which they dream not of.
     Henry V II, ii
     [*Shakespeare, Secret Intelligence and Statecraft* 1962, W F Friedman *Am. Phil. Soc.,*
     Philadelphia]

61   For speculation turns not to itself
     For it hath travell'd and is married there
     Where it may view itself.
     Troilus and Cressida III, iii

62   With thy sharp teeth this knot intrinsicate
     Of life at once untie;
     Anthony and Cleopatra V, ii

63   And as imagination bodies forth
     The form of things unknown, and the poet's pen
     Turns them to shapes, and gives to airy nothing

A local habitation and a name.
A Midsummer Night's Dream V, i

64   quasi person
Love's Labour Lost IV, ii

65   We have our philosophical persons, to make modern and familiar, things
supernatural and causeless.
All's Well That Ends Well II, iii

66   These are begot in the ventricle of memory, nourished in the womb of *pia
mater.*
Love's Labour Lost

Karl Shapiro 1913–

67   Who loved the ugly beauty of a painted cube.
Trial of a Poet 1947 (New York: Reynal & Hitchcock)

Roger Shattuck 1923–

68   Pataphysics is the science of the particular, of laws governing exceptions. . ..
Pataphysics is the science of imaginary solutions.
[Of the Dada-ist art movement]
Au seuil de pataphysique, College de Pataphysique, near Rue de Seine, Paris. In C H
Waddington *Beyond Appearance* 1969 (Edinburgh University Press)

George Bernard Shaw 1856–1950

69   All problems are finally scientific problems.
In Preface to *The Doctor's Dilemma* from *Collected Prefaces* by Constable

70   The fact that a believer is happier than a sceptic is no more to the point
than the fact that a drunken man is happier than a sober one. The happiness
of credulity is a cheap and dangerous quality.
In Preface to *Androcles and the Lion*

71   Getting patronage is the whole art of life. A man cannot have a career
without it.
Captain Brassbound's Conversion 1906, Act III

72   Great art is never produced for its own sake. It is too difficult to be worth
the effort.
In Preface to *Three Plays by Brieux*

73   Tyndall declared that he saw in Matter the promise and potency of all
forms of life, and with his Irish graphic lucidity made a picture of a world
of magnetic atoms, each atom with a positive and negative pole, arranging
itself by attraction and repulsion in orderly crystalline structure. Such a
picture is dangerously fascinating to thinkers oppressed by the bloody disorders of the living world. Craving for purer subjects of thought, they find
in the contemplation of crystals and magnets a happiness more dramatic
and less childish than the happiness found by mathematicians in abstract
numbers, because they see in the crystals beauty and movement without
the corrupting appetites of fleshly vitality.
In Preface to *Back to Methuselah* 1921

74   'Very nice sort of place, Oxford, I should think, for people that like that
sort of place. They teach you to be a gentleman there. In the polytechnic
they teach you to be an engineer or such like. See?'
Straker, in *Man and Superman*, Act 2

75 The fact that doctors themselves die of the very diseases they profess to cure passes unnoticed.
In the Preface to *The Doctor's Dilemma* 1911

76 'I've got a soul: dont tell me I havnt. Cut me up and you cant find it. Cut up a steam engine and you cant find the steam. But, by George, it makes the engine go.'
[Shaw liked his own orthography]
Tarleton in *Misalliance*

77 The right to knowledge is not the only right; and its exercise must be limited by respect for other rights, and for its own exercise by others. ...so the wisest people say, 'If you cannot attain to knowledge without torturing a dog, you must do without knowledge.'
The Doctor's Dilemma, Preface, 1911

[Bishop] Fulton Sheen 1895–

78 An atheist is a man who has no invisible means of support.
Look 14 December 1955

Charles S Sheldon

79 What is the long-run psychological cost to us of having the backside of the Moon dotted with Soviet names? Will they do the same for Mars? To pretend that national prestige is unimportant is to show a limited awareness of historical forces in society.
National Goals in Space, NASA in *Proceedings of the Working Conference on Space Nutrition and Related Waste Problems, Tampa, Florida* 27 April 1964

Percy Bysshe Shelley 1792–1822

80 ...happiness
And science dawn through late upon the earth;...
Queen Mab VIII, 11.227–8

81 The gigantic shadows which futurity casts upon the present.
The Defence of Poetry

82 He gave man speech, and speech created thought,
Which is the measure of the universe;
And Science struck the thrones of earth and heaven.
Prometheus Unbound 1820

83 Power, like a desolating pestilence,
Pollutes what e'er it touches; and obedience,
Bane of all genius, virtue, freedom, truth,
Makes slaves of men, and, of the human frame,
A mechanical automaton.

Shen Kua 1031–1095

84 Salt is a means to wealth, profit without end emerging from the sea.
Hsu tsu chih t'ung chien ch'ang pien. 280: 17b–21b. *Dictionary of Scientific Biography*, N Sivin, 1975

[Sir] Charles S Sherrington 1857–1952

85   Like this old Earth that lolls through sun and shade,
      Our part is less to make than to be made.
      Man on his Nature 1955 (Harmondsworth: Penguin)

Ernst Werner von Siemens 1816–1892

86   *Messen ist wissen.*
      To measure is to know.

Herbert Alexander Simon 1916–

87   A man, viewed as a behaving system, is quite simple. The apparent com-
      plexity of his behaviour over time is largely a reflection of the complexity
      of the environment in which he finds himself.
      The Sciences of the Artificial 1969 (Cambridge, MA: The MIT Press)

[Sir] John Sinclair 1754–1835

88   At present there are a greater number of intelligent practical chemists in
      Scotland, in proportion to the population, than perhaps in any other coun-
      try in the world.
      1814

R G H Siu

89   The blackmailing of intellectual freedom by expensive, specialized, fixed
      installations is one of the dangers of the current financial plenty for
      research...an apparatus of considerable cost is built to carry out certain un-
      usual measurements, only to doom the servile scientists to a life of repetitive
      measurements. Soon the machine assumes the dictatorship of research....
      It is a puzzle...that scientists so outspokenly against thought control by
      other men would so unwittingly embrace thought control by money and
      machines.
      The Tao of Science 1957 (Cambridge, MA: MIT Press)

Burrhus Frederic Skinner 1904–1990

90   'The science of behaviour is full of special twists like that,' said Frazier.
      'It's the science of science a special discipline concerned with talking about
      talking and knowing about knowing. Well, there's a motivational twist too.
      Science in general emerged from a competitive culture. Most scientists are
      still inspired by competition or at least supported by those who are. But
      when you come to apply the methods of science to the special study of
      human behaviour, the competitive spirit commits suicide. It discovers the
      extraordinary fact that in order to survive, we must in the last analysis,
      not compete.'
      Walden Two 1948 (New York: Macmillan)

91   Education is what survives when what has been learnt has been forgotten.
      New Scientist 21 May 1964

92   The real problem is not whether machines think but whether men do.
      Contingencies of Reinforcement 1969. Chapter 9

Martyn Skinner 1906–

93   For we are like the Chinese in reverse.
      Our feeling for the future's so prodigious
      We might be termed descendant worshippers.
      The Return of Arthur 1966 (London: Chapman & Hall)

Herman Skolnik 1914–

94   Science and engineering students [presumably] are left to learn about their
      literature in the same way they learn about sex.
      J. Chem. Inf. Comput. Sci. 1981 **21** No 4 174

[Air Chief Marshal] Sir John Slessor 1897–

95   A few years ago it would never have occurred to me, or, I think, to any offi-
      cer of any fighting service, that what the RAF soon came to call a "Boffin",
      a gentleman in grey flannel bags, whose occupation in life had previously
      been something markedly unmilitary such as biology or physiology, would
      be able to teach us a great deal about our business. Yet so it was.
      In B Johnson *The Secret War* BBC, 1978. 1979 (London: Arrow)

Samuel Smiles 1812–1904

96   We often discover what will do, by finding out what will not do; and prob-
      ably he who never made a mistake never made a discovery.
      Self-help Chapter 11

Adam Smith 1723–1790

97   The machines that are first invented to perform any particular movement
      are always the most complex, and succeeding artists generally discover
      that with fewer wheels, with fewer principles of motion than had originally
      been employed, the same effects may be more easily produced. The first
      philosophical systems, in the same manner, are always the most complex.
      Essay on the Principles which Lead and Direct Philosophical Inquiries

98   In opulent and commercial societies, besides, to think or to reason comes
      to be, like every other employment, a particular business, which is carried
      on by a very few people, who furnish the public with all the thought and
      reason possessed by the vast multitudes that labour.
      In W R Scott *Adam Smith as Student and Professor* 1937

99   [Only a very small part of any ordinary person's knowledge has been the
      produce of his own observation or reflection] all the rest has been the pur-
      chased, in the same manner as his shoes or his stockings, from those whose
      business is to make up and prepare for market that particular species of
      goods.
      In W R Scott *Adam Smith as Student and Professor* 1937

100   Science is the great antidote to the poison of enthusiasm and superstition.
      The Wealth of Nations, 5.3, 1776

Cyril Stanley Smith 1903–

101   Matter is a holograph of itself in its own internal radiation.
      Letter to A L Mackay, 1968

John Maynard Smith 1920–

102   [On J B S Haldane's book *Possible Worlds*]
The mixture of intellect and blasphemy was absolutely overwhelming and
I've been attracted to that all the rest of my life.
A Passion for Science ed L Wolpert and A Richards 1988 (Oxford: Oxford University
Press)

Sydney Smith 1771–1845

103   Science is his forte, and omniscience his foible.
[Of William Whewell]
In Issac Todhunter *William Whewell* (Farnborough: Gregg International)

[General] Walter Bedell Smith 1895–1961

104   My big job is to get the best brains in the country, persuade them to leave
fame and fortune for a government job where they'll study secrets they
can't even discuss with their wives...
[Second Director of the CIA]
They Call it Intelligence 1963 (London: Abelard–Schuman)

[Lord] Charles Percy Snow 1905–1980

105   A good many times I have been present at gatherings of people who, by the
standards of traditional culture, are thought highly educated and who have
with considerable gusto been expressing their incredulity at the illiteracy of
scientists. Once or twice I have been provoked and have asked the company
how many of them could describe the Second Law of Thermodynamics. The
response was cold: it was also negative.
The Two Cultures The Rede Lecture, 1959 (London: Cambridge University Press)

106   The scientific revolution is the only method by which most people can
gain the primal things (years of life, freedom from hunger, survival for
children)—the primal things which we take for granted and which have in
reality come to us through having had our own scientific revolution not so
long ago.
The Two Cultures: A Second Look 1963 (London: Cambridge University Press)

107   Scientists have it within them to know what a future-directed society feels
like, for science itself, in its human aspect, is just like that.
A Postscript to Science and Government 1962 (Oxford: Oxford University Press)

108   Scientists.... I should say that naturally they had the future in their bones.
The Two Cultures The Rede Lecture, 1959 (London: Cambridge University Press)

109   [Of molecular biology] This branch of science is likely to affect the way in
which men think of themselves more profoundly than any scientific advance
since Darwin's and probably more so than Darwin's.
The Two Cultures: A Second Look 1963 (London: Cambridge University Press)

110   We cannot delude ourselves that this new invention [the atomic bomb] will
be better used. Yet it must be made, if it really is a physical possibility. If
it is not made in America this year, it may be next year in Germany. There
is no ethical problem; if the invention is not prevented by physical laws, it
will certainly be carried out somewhere in the world. It is better, at any
rate, that America should have six months start.
But again, we must not pretend. Such an invention will never be kept

secret; the physical principles are too obvious, and within a year every big laboratory on earth would have come to the same result. For a short time, perhaps, the U.S. Government may have this power entrusted to it; but soon after it will be in less civilised hands.
Editorial in *Discovery* 1939

111   [The scientist narrator commenting on politicians attending a party.]
'These people's politics were not my politics. They didn't know the world they were living in, much less the world that was going to come.'
Corridors of Power 1964

Socrates 470–399 BC

112   It seemed to me a superlative thing to know the explanation of everything, why it comes to be, why it perishes, why it is.
Plato *Phaedo*, 96a. Put into Socrates' mouth.

Frederick Soddy 1877–1956

113   Four circles to the kissing come,
The smaller are the benter.
The bend is just the inverse of
The distance from the centre.
Though their intrigue left Euclid dumb
There's now no need for rule of thumb.
Since zero bend's a dead straight line
And concave bends have minus sign,
The sum of square of all four bends
Is half the square of their sum.
Nature 1936 **137** 1021
[The Problem of Apollonius]

Omond McKillop Solandt 1909–

114   It is presumptuous for scientists to try to formulate national goals, since science is by no means the only important activity in the nation. But scientists have a duty to point out that most nations have neither explicit goals nor a mechanism for formulating them.
[Chairman of the Science Council of Canada]

Sophocles 495–406 BC

115   One must learn by doing the thing; though you think you know it, you have no certainty until you try.
Trachiniae 592

Albert Speer 1905–1981

116   The fact that young scientists were also important was not discovered until 1942.
Inside the Third Reich 1975 (New York: Sphere Books). ©Macmillan (New York) 1970

117   We owed the success of our programs to thousands of technicians with special achievments to their credit to whom we now entrusted the responsibility for whole segments of the armaments industry. This aroused their buried enthusiasm. They also took gladly to my unorthodox style of leadership. Basically, I exploited the phenomenon of the technician's often blind devotion to his task.
Inside the Third Reich 1975 (New York: Sphere Books). ©Macmillan (New York) 1970

Herbert Spencer 1820–1903

118   Definition of life: The continuous adjustment of internal relations to external relations.
Principles of Biology section 30

119   Science is organised knowledge.
Education Chapter 29

120   Knowledge of the lowest kind is un-unified knowledge. Science is partly unified knowledge. Philosophy is completely unified knowledge.

121   The general law of organisation...is that distinct duties [=functions] entail distinct structures.
Oxford English Dictionary 1959

122   Integration keeps pace with differentiation.
In C G Phillips *PRS* 1977 **B199** 415–24

123   Survival of the fittest.
Principles of Biology III, Chapter 12, section 164

Stephen Spender 1909–

124   More beautiful and soft than any moth
With burring furred antennæ feeling its huge path
Through dusk, the airliner with shut-off engines
Glides over suburbs and the sleeves set trailing tall
To point the wind. Gently, broadly she falls,
Scarcely disturbing charted currents of air.
The Landscape near an Aerodrome in *Collected Poems, 1928–53* (London: Faber & Faber)

Oswald Spengler 1880–1936

125   Nature is the shape in which the man of higher cultures synthesizes and interprets the immediate impressions of his senses. History is that from which his imagination seeks comprehension of the living existence of the world.
The Decline of the West transl C F Atkinson, vol 1, 1926 (London: Allen & Unwin)

Edmund Spenser ca 1552–1599

126   Within this wide great Universe
Nothing doth firme and permanent appeare,
But all things tost and turned by transverse.
Two Cantos of Mutability 1609

Thomas Sprat 1635–1713

127   Invention is a heroic thing and placed above the reach of a low and vulgar genius. It requires an active, a bold, a nimble, a restless mind: a thousand difficulties must be contemned with which a mean heart would be broken: many attempts must be made to no purpose: much treasure must be scattered without any return: much violence and vigour of thought must attend it: some irregularities and excesses must be granted that could hardly be pardoned by the severe rules of prudence.
The History of the Royal Society 1667

128 A close, naked, natural way of speaking; positive expressions, clear senses;
a native easiness; bringing all things as near the Mathematical plainness,
as they can: and preferring the language of the Artisans, Countrymen, and
Merchants before that of Wits and Scholars.
The History of the Royal Society 1667

Josef Vissarionovich Stalin 1879–1953

129 The writer is the engineer of the human soul.

130 [On the telephone]
It will unmake our work. No greater instrument of counter-revolution and
conspiracy can be imagined.
L D Trotsky *Life of Stalin*

131 Only idealists can speak of thinking without language.
On Linguistics. Nature 1982 **299** 377

132 We Soviet people have not a little experience with the technical
intelligentsia...
In Lovett Dickson *H G Wells*

Barbara Stanwyck (Ruby Stevens) 1907–

133 [During a trans-Atlantic crossing on the Queen Mary]
What time does this place get to New York?

State of Tennessee, USA

134 It shall be unlawful for any teacher in any of the universities, normals and
all other public schools of the state which are supported in whole or in part
by the public school funds of the state, to teach any theory that denies the
story of the divine creation of man as taught in the Bible, and to teach
instead that man has descended from a lower order of animals.
1925. Repealed 1967

Gertrude Stein 1874 1946

135 When we were having a book printed in France we complained about the
bad alignment. Ah, they explained, that is because they use machines now,
machines are bound to be inaccurate, they have not the intelligence of
human beings, naturally the human mind corrects the fault of the hand,
but with a machine, of course, there are errors. The reason why all of us
naturally began to live in France is because France has scientific methods,
machines and electricity, but does not really believe that these things have
anything to do with the real business of living.
Paris France 1940 (New York: Scribners)

136 [Last words]
What is the answer? [pause] What is the question?

Charles Proteus Steinmetz 1865–1923

137 There are no foolish questions and no man becomes a fool until he has
stopped asking questions.

Stendhal [Henri Beyle] 1783–1842

138 *Ce que j'appelle cristallisation, c'est opération de l'esprit, qui tire de tout
ce qui se présente la découverte que l'objet aimé a des nouvelles perfections.*

I call 'crystallisation' that action of the mind that discovers fresh perfections
in its beloved at every turn of events.
De l'amour 1822, Chapter 1

Gunther Stent 1924–

139   It would, of course, be a poor lookout for the advancement of science if
young men started believing what their elders tell them, but perhaps it is
legitimate to remark that young Turks look younger, or more Turkish,...if
the conclusions they eventually reach are different from what anyone had
said before.
Nature 1969 **221** 320

140   I believe that science is, by nature, reductionist, but I also believe that
reductionism will not carry us all the way. One of the reasons why I think
science will eventually peter out is because you must always explain some
higher level in terms of some lower level—that's what scientists have to do.
But I think that when finally we get to sufficiently complex things, this
will not be possible. It is precisely because I think reductionism will have
to fail, that I believe that science is coming to an end.
A Passion for Science ed L Wolpert and A Richards 1988 (Oxford: Oxford University
Press)

George Stephenson 1781–1848

141   I will send them the locomotive to be the Great Missionary among them.

Laurence Sterne 1713–1768

142   It is in the nature of a hypothesis when once a man has conceived it, that
it assimilates everything to itself, as proper nourishment, and from the first
moment of your begetting it, it generally grows stronger by everything you
see, hear or understand.
Tristram Shandy 1759–1767

143   Through prolonged close contact and friction with the objects of their study,
the minds of experts finally acquire a pictorial, moth-like, fiddling perfection.

Wallace Stevens 1879–1955

144   The man bent over his guitar,
A shearsman of sorts. The day was green.
They said, 'You have a blue guitar,
You do not play things as they are'.
The man replied, 'Things as they are
Are changed upon a blue guitar...
The Man with the Blue Guitar in *Collected Poems of Wallace Stevens* 1937 (London:
Faber & Faber)

Adlai E Stevenson 1900–1965

145   *Via ovicipitum dura est.*
The way of the egghead is hard.
[Attributed]

146   We travel together, passengers in a little space-ship, dependent on its vul-
nerable supplies of air and soil.
Speech to the UN Economic and Social Council, 1965. In J Maddox *The Doomsday
Syndrome* 1972 (London: Macmillan)

Strabo 1st Century BC

147 The poets were not alone in sponsoring myths. Long before them cities and
lawmakers had found them a useful expedient.... They needed to control
the people by superstitious fears, and these cannot be aroused without
myths and marvels.
Geography I, 2; 8

Alexander Strange 1818–1876

148 1. That science is essential to the advancement of civilisation, the development of national wealth, and the maintenance of national power.
2. That all science should be cultivated, even branches of science which do
not appear to promise immediately direct advantage.
3. That the State or Government, acting as trustees of the people, should
provide for the cultivation of those departments of science which, by reason
of costliness, either in time or money, or of remoteness of probable profit,
are beyond the reach of private individuals; in order that the community
may not suffer from the effect of insufficiency of isolated effort.
4. That to whatever extent science may be advanced by State agency, that
agency should be systematically constituted and directed.
*Conclusions of Devonshire Royal Commission on Scientific Institutions and the Ad-
vancement of Science* 1872

Igor Stravinsky 1882–1973

149   The more constraints one imposes, the more one frees one's self of the chains that shackle the spirit...the arbitrariness of the constraint only serves to obtain precision of execution.
Poetics of Music (Cambridge, MA: Harvard University Press)

Johann August Strindberg 1849–1912

150   Then is it reasonable to think that one can see, by looking in a microscope, what is going on in another planet?
[By the examination of meteorites]
The Father 1887, Act 1, scene 5

151   A generation that had the courage to get rid of God, to crush the state and church, and to overthrow society and morality, still bowed before Science. And in Science, where freedom ought to reign, the order of the day was 'believe in the authorities or off with your head'.
Antibarbarus

A M Sullivan 1896–

152   Take Carbon for example then
What shapely towers it constructs
To house the hopes of men!
Atomic Architecture

James B Sumner 1887–1955

153   Nothing can take the place of intelligence.
TIBS, July 1981

Sun Tze 5th–6th Century BC

154   Hence to fight and conquer in all your battles is not supreme excellence; supreme excellence consists in breaking the enemy's resistance without fighting.
[Favourite author of Mao Tse-tung, myself and many others]
Sun Tze Ping Fa transl L Giles, 1910 (London: Luzac)

155   [Sun the Master said:] The control of large numbers is possible, and like unto that of small numbers, if we subdivide them.
Sun Tze Ping Fa Chapter 5

156   In war there are but two forces, the expected and the unexpected; but they are capable of infinite variation. Their mutual interchange is like a wheel, having neither beginning nor end. They are a mystery that none can penetrate.
Sun Tze Ping Fa Chapter 5

Sun Yat Sen 1866–1925

157   [In China under the Manchu Dynasty...]
No one is allowed, on pain of death, to invent anything new, or to make known any new discovery.
Kidnapped in London 1897, London

Jonathan Swift 1667–1745

158 If they would, for example, praise the beauty of a woman, or any other animal. they describe it by rhombs, circles, parallelograms, ellipses, and other geometrical terms...
A Voyage to Laputa in *Gulliver's Travels*

159 In the school of political projectors, I was but ill entertained, the professors appearing, in my judgment, wholly out of their senses; which is a scene that never fails to make me melancholy. These unhappy people were proposing schemes for persuading monarchs to choose favourites upon the score of their wisdom, capacity, and virtue; of teaching ministers to consult the public good; of rewarding merit, great abilities, and eminent services; of instructing princes to know their true interest, by placing it on the same foundation with that of their people; of choosing for employment persons qualified to exercise them; with many other wild impossible chimeras, that never entered before into the heart of man to conceive, and confirmed in me the old observation, that there is nothing so extravagant and irrational which some philosphers have not maintained for truth.
A Voyage to Laputa in *Gulliver's Travels*

160 ...whoever could make two ears of corn, or two blades of grass, to grow upon a spot of ground where only one grew before, would deserve better of mankind, and do more essential service to this country, than the whole race of politicians put together.
A Voyage to Brobdingnag in *Gulliver's Travels*

161 [Of the Laputans] They have likewise discovered two lesser stars, or satellites, which revolve about Mars, whereof the innermost is distant from the centre of the primary planet exactly three of his diameters, and the outermost five; the former revolves in the space of ten hours, and the latter in twenty one and a half;...
[These satellites were first observed by Asaph Hall in 1877. Their periods are 7.7 and 30 hours]
A Voyage to Laputa in *Gulliver's Travels*

John Lighton Synge 1897–

162 [Irish mathematician]
'The northern ocean is beautiful', said the Orc, 'and beautiful the delicate intricacy of the snowflake before it melts and perishes, but such beauties are as nothing to him who delights in numbers, spurning alike the wild irrationality of life and the baffling complexity of nature's laws.'
Kandelman's Krim 1957 (London: Cape)

Albert Szent-Györgyi 1893–1986

163   Discovery consists of seeing what everybody has seen thinking what nobody has thought.
The Scientist Speculates ed I G Good, 1962 (London: Heinemann)

164   It is common knowledge that the ultimate source of all our energy and negative entropy is the radiation of the sun. When a photon interacts with a material particle on our globe it lifts one electron from an electron pair to a higher level. This excited state as a rule has but a short lifetime and the electron drops back within 10^{-7} to 10^{-8} seconds to the ground state giving off its excess energy in one way or another. Life has learned to catch the electron in the excited state, uncouple it from its partner and let it drop back to the ground state through its biological machinery utilizing its excess energy for life processes.
Light and Life ed W D McElroy and B Glass, 1961 (Baltimore, MD: Johns Hopkins Press)

165   Knowledge is a sacred cow, and my problem will be how we can milk her while keeping clear of her horns.
Science 1964 **146** 1278

166   The real scientist...is ready to bear privations and, if need be, starvation rather than let anyone dictate to him which direction his work must take.
Science Needs Freedom 1943

167   Research means going out into the unknown with the hope of finding something new to bring home. If you know in advance what you are going to do, or even to find there, then it is not research at all: then it is only a kind of honourable occupation.
Perspectives in Biology and Medicine 1971

168   The greatest stride in biology, in our century, was its shift to the molecular dimension. The next will be its shift toward the sub-molecular, electronic dimension.
Bioelectronics 1968 (New York: Academic)

169   If you would ask a chemist to find out for you what a dynamo is, the first thing he would do is to dissolve it in hydrochloric acid. A molecular biochemist would, probably, take the dynamo to pieces, describing carefully the helices of wire. Should you timidly suggest to him that what is driving the machinery may be, perhaps, an invisible fluid, electricity, flowing through.

170   Nature is built on large principles. Nature does not build distinct principles for a tree, a bush, a flower or a man. All are built on a common, large basic principle—therefore it does not really matter what is the topic that we study, if we are intelligent enough to understand the basic principle, the structure of life.

171   [A vitamin is] a substance that makes you ill if you don't eat it.

Leo Szilard 1898–1964

172   Don't lie if you don't have to.
Science 1972 **176** 966

173 If you want to succeed in this world you don't have to be much cleverer than other people, you just have to be one day earlier.
In S R Weart and G W Szilard (ed) *Leo Szilard: His Version of the Facts* 1978 (Cambridge, MA: MIT Press)

174 I have been asked whether I would agree that the tragedy of the scientist is that he is able to bring about great advances in our knowledge, which mankind may then proceed to use for purposes of destruction. My answer is that this is not the tragedy of the scientist; it is the tragedy of mankind.
In S R Weart and G W Szilard (ed) *Leo Szilard: His Version of the Facts* 1978 (Cambridge, MA: MIT Press)

Taibai died 1842

1   To a man who has not eaten a globe-fish, we cannot speak of its flavour.
[This is the *fugu*, parts of which are extremely poisonous containing tetrodotoxin]
In R H Blyth *Zen in English Literature* 1942 (Tokyo: Hokuseido Press)

Hippolyte Taine 1828–1893

2   *Le vice et la vertu sont des produits comme le vitriol et le sucre.*
Vice and virtue are products like sulphuric acid and sugar.
Histoire de la littérature anglaise 1863, Introduction

3   *On peut assez exactement comparer l'intérieur d'une tête anglaise à un guide de Murray: beaucoup de faits et peu d'idées.*
You could compare the inside of an Englishman's head exactly to one of John Murray's guides: stuffed with facts but containing few ideas.
Notes sur Angleterre 1872

Peter Guthrie Tait 1831–1901

4   Perphaps to the student there is no part of elementary mathematics so repulsive as is spherical trigonometry.
Encyclopaedia Britannica 1911, 11th edn, 22, 721 (article on *Quaternions*)

Charles-Maurice De Talleyrand 1745–1838

5   Both erudition and agriculture ought to be encouraged by government; wit and manufactures will come of themselves.

6   It is sometimes quite enough for a man to feign ignorance of that which he knows, to gain the reputation of knowing that of which he is ignorant.

Vasili Nikitich Tatishchev 1686–1750

7   Freedom is not an essential and basic condition for the growth of science; the care and diligence of government authorities are the most important conditions for this development.
Razgovor
In A Vacinich *Science in Russian Culture* 1965 (London: Peter Owen)

W P Taylor

8   It is somewhat disturbing to realise, however, that Western civilisation survives only by virtue of the inefficiency of its military establishments, and that many of our ablest scientists are devoting their lives to eliminating that inefficiency.
Bul. Atomic Scientists September 1976 **32** 2

Alfred [Lord] Tennyson 1809–1892

9   ...Saw the heavens fill with commerce, argosies of magic sails,
Pilots of the purple twilight, dropping down with costly bales;
Heard the heavens fill with shouting, and there rained a ghastly dew
From the nations' airy navies grappling in the central blue....

10   Science moves, but slowly slowly, creeping on from point to point
Slowly comes a hungry people, as a lion creeping nigher,
Glares at one that nods and winks behind a slowly-dying fire.

11 ...Better fifty years of Europe than a cycle of Cathay.
    [Prophetic—which excuses the ghastly rhyme]
    [A cycle of Cathay is sixty years]
    Locksley Hall 1832

12 Science grows and Beauty dwindles.
    Locksley Hall Sixty Years After 1986

13 A time to sicken and to swoon,
    When Science reaches forth her arms
    To feel from world to world, and charms
    Her secret from the latest moon.
    In Memoriam AHH XXI, 4

14 ...yearning in desire
    To follow knowledge, like a sinking star
    Beyond the utmost bound of human thought...
    ...
    To strive, to seek to find and not to yield.
    Ulysses

Tertullian ca 155–222

15 *Certum est quia imposibile est.*
    It is certain because it is impossible.
    [*Or*: It is so extraordinary that it must be true]
    De Carne Cristi 5

Thales of Miletus ca 640–546 BC

16 Water is the principle, or the element, of things.
    All things are water.
    Plutarch *Placita Philosophorum* i, 3

Margaret Hilda Thatcher 1925–

17 [During the Falklands war, 1982]
    When you've spent half your political life dealing with humdrum issues like
    the environment...it's exciting to have a real crisis on your hands.

Stefan Themerson 1910–

18 With a poet it's different,
    If his poem is bad
    Even his broken heart
    Will not make it better.
    On Semantic Poetry 1975 (London: Gaberbocchus)

Theophrastus ca 370–ca 287 BC

19 It is manifest that Art imitates Nature and sometimes produces very pe-
    culiar things.
    Theophrastus' History of Stones transl J Hill, 1746 (London)

René Thom 1923–

20 'A l'heure actuelle la Biologie n'est qu'un immense cimitière de faits, vague-
    ment synthesés par un petit nombre de formules creusés.

> At present biology is nothing more than an immense graveyard of facts,
> vaguely held together by a few empty formulas.
> 1974. In E C Zeeman (ed) *Catastrophe Theory* 1977 (Reading, MA: Addison-Wesley)

Dylan Thomas 1914–1953

21   Desire is phosphorus: the chemic bruit.
      To Leda

22   Stretch the salt photographs.
      My Neophyte

Lewis Thomas 1913–

23   What is [the Earth] most like?...It is most like a single cell.
      The Lives of a Cell 1974

24   Music is the effort we make to explain to ourselves how our brains
      work...*The Art of Fugue* is not a special pattern of thinking.... The whole
      piece is not about thinking about something, it is about thinking. If you
      want, as a experiment, to hear the whole mind working, all at once, put on
      the St. Matthew Passion and turn the volume up all the way. That is the
      sound of the whole central nervous system of human beings, all at once.
      The Medusa and the Snail 1981 (London: Penguin)

R S Thomas 1913–

25   God woke, but the nightmare
      Did not recede. Word by word
      The tower of speech grew.
      He looked at it from the air
      He reclined on. One word more and
      It would be on a level with him; vocabulary
      Would have triumphed.... He leaned
      Over and looked at the dictionary
      They used. There was the blank still
      By his name...
      The Gap from *Frequencies* 1978 (London: Macmillan)

D'Arcy Wentworth Thompson 1860–1948

26   Cell and tissue, shell and bone, leaf and flower, are so many portions of
      matter, and it is in obedience to the laws of physics that their particles
      have been moved, moulded and confirmed. They are no exception to the
      rule that God always geometrizes. Their problems of form are in the first
      instance mathematical problems, their problems of growth are essentially
      physical problems, and the morphologist is, *ipso facto*, a student of physical
      science.
      On Growth and Form 1917 (London: Cambridge University Press)

27   Form is a diagram of forces.
      On Growth and Form 1917 (London: Cambridge University Press)

28   It behoves us always to remember that in physics it has taken great men to
      discover simple things. They are very great names indeed which we couple
      with the explanation of the path of a stone, the droop of a chain, the tints
      of a bubble, the shadows in a cup.
      On Growth and Form 1917 (London: Cambridge University Press)

29 ...chemistry, the most cosmopolitan of sciences, the most secret of arts.
The Legacy of Greeces 1921 (Oxford: Oxford University Press)

Francis Thompson 1859–1907

30 All things by immortal power,
Near and far
Hiddenly
To each other linked are,
That thou canst not stir a flower
Without troubling of a star.
The Mistress of Vision

W H Thompson 1810–1886

31 [Master of Trinity College, Cambridge]
We are none of us infallible: not even the youngest of us.
G M Trevelyan *Trinity College* 1943 (Cambridge: Cambridge University Press)

James Thomson 1700–1748

32 Even now the setting sun and shifting clouds,
Seen, Greenwich, from thy lovely heights, declare
How just, how beauteous the refractive law.
To the Memory of Sir Isaac Newton 1727

33 [Newton] from motion's simple laws
Could trace the secret hand of Providence,
Wide-working through this universal frame.
To the Memory of Sir Isaac Newton 1727

[Sir] Joseph John Thomson 1856–1940

34 This example illustrates the differences in the effects which may be produced by research in pure or applied science. A research on the lines of applied science would doubtless have led to improvement and development of the older methods—the research in pure science has given us an entirely new and much more powerful method. In fact, research in applied science leads to reforms, research in pure science leads to revolutions, and revolutions, whether political or industrial, are exceedingly profitable things if you are on the winning side.
In Lord Rayleigh *J J Thomson* 1943 (London: Cambridge University Press)

William Thomson [Lord Kelvin] 1824–1907

35 Do not imagine that mathematics is hard and crabbed, and repulsive to common sense. It is merely the etherialization of common sense.
In S P Thompson *Life of Lord Kelvin* 1910

36 Fourier is a mathematical poem.
W Thomson and P G Tait *Treatise on Natural Philosophy* vol 1, pp 713, 718. Quoted by F Engels in *The Dialectics of Nature*

37 I am never content until I have constructed a mechanical model of the subject I am studying. If I succeed in making one, I understand; otherwise I do not.
Notes of Lectures on *Molecular Dynamics and the Wave Theory of Light*

38   I often say that when you can measure what you are speaking about, and
      express it in numbers, you know something about it; but when you cannot
      measure it, when you cannot express it in numbers, your knowledge is of a
      meagre and unsatisfactory kind.
      Lecture to the Institution of Civil Engineers, 3 May 1883

39   [Of the ether] it is no greater mystery at all events than the shoemaker's
      wax.
      [But in fact Carnauba wax is very complicated furnishing, *inter alia*, the first example
      of an electret]
      In J A Lenihan *Science in Action* 1979 (Bristol: Institute of Physics)

40   Scientific wealth tends to accumulate according to the law of compound
      interest. Every addition to knowledge of the properties of matter supplies
      the [physical scientist] with new instrumental means for discovering and
      interpreting phenomena of nature, which in their turn afford foundations
      of fresh generalisations, bringing gains of permanent value into the great
      storehouse of [natural] philosophy.
      British Association, Presidential address, 1871

Henry David Thoreau 1817–1862

41   It appears to be law that you cannot have a deep sympathy with both man
      and nature.
      Walden

42   Simplicity, simplicity, simplicity. I say, let your affairs be as two or three,
      and not a hundred or a thousand; instead of a million count half a dozen,
      and keep your accounts on your thumb nail.
      Walden. See *Scientific American* August 1969

43   Time is but the stream I go fishing in.

Thucydides ca 457–399 BC

44   *Pericles*: Our city is thrown open to the world, and we never expel a for-
      eigner or prevent him from seeing or learning anything of which the secret
      if revealed to an enemy might profit him. We do not rely upon management
      or trickery, but upon our own hearts and hands.
      [But Athens lost the war!]
      R V Jones *Most Secret War* 1978 (London: Hamish Hamilton)

James Thurber 1894–1961

45   Progress was all right; it only went on too long.
      [Attributed]

[Archbishop] John Tillotson 1603–1694

46   How often might a man, after he had jumbled a set of letters in a bag, fling
      them out upon the ground before they would fall into an exact poem, yea,
      or so much as make a good discourse in prose! And may not a little book
      be as easily made by chance as this great volume of the world?
      In C S Pierce *Philosophical Writings of Pierce* 1955 (New York: Dover)

Kliment Arkadievich Timiryazev 1843–1920

47   I set myself two parallel tasks: to create for science and write for the people.
      Science and Democracy 1927 (Leningrad: Priboi)

[Sir] Henry Tizard 1885–1959

48 Andrade [who was looking after wartime inventions] is like an inverted Micawber, waiting for something to turn down.
In C P Snow *A Postscript to Science and Government* 1962 (Oxford: Oxford University Press)

49 The secret of science is to ask the right question, and it is the choice of problem more than anything else that marks the man of genius in the scientific world.
In C P Snow *A Postscript to Science and Government* 1962 (Oxford: Oxford University Press)

James Todd

50 Sowai Jey Sing [Jai Singh—Rajah of Jaipur and constructor of the great observatory of Jaipur] died in S1709 [1743 AD] having ruled forty four years. Three of his wives and several concubines ascended his funeral pyre, on which Science expired with him.
Annals and Antiquities of Rajasthan 1829–1957 **2**, 298 (London: Routledge)

Isaac Todhunter 1820–1884

51 [Mathematician, of St. John's College, Cambridge.]
[Maxwell asked him whether he would like to see an experimental demonstration of conical refraction.] No. [he replied] I have been teaching it all my life, and I do not want to have my ideas upset.
J G Crowther *The Statesmen of Science* 1965 (London: Cresset Press)

[Count] Lev Nikolaevich Tolstoi 1828–1910

52 I am convinced that the history of so-called scientific work in our famous centuries of European civilisation will, in a couple of hundred years, represent an inexhaustible source of laughter and sorrow for future generations. The learned men of the small western part of our European continent lived for several centuries under the illusion that the eternal blessed life was the West's future. They were interested in the problem of when and where this blessed life would come. But they never thought of how they were going to make their life better.
1884. Probably in *What is Religion?*

53 Our body is a machine for living.
Napoleon in *War and Peace* transl L and A Maude, Book X, 1922 (Oxford: Oxford University Press) Chapter 29

54 A modern branch of mathematics, having achieved the art of dealing with the infinitely small, can now yield solutions in other more complex problems of motion, which used to appear insoluble.
This modern branch of mathematics, unknown to the ancients, when dealing with problems of motion, admits the conception of the infinitely small, and so conforms to the chief condition of motion (absolute continuity) and thereby corrects the inevitable error which the human mind cannot avoid when dealing with separate elements of motion instead of examining continuous motions.
In seeking the laws of historical movement just the same thing happens. The movement of humanity, arising as it does from innumerable human

wills, is continuous.

To understand the laws of this continuous movement is the aim of history....

Only by taking an infinitesimally small unit for observation (the differential of history, that is, the individual tendencies of men) and attaining to the art of integrating them (that is, finding the sum of these infinitesimals) can we hope to arrive at the laws of history.

[The molecular dynamics approach to history]

War and Peace transl L and A Maude, Book XI, 1922 (Oxford: Oxford University Press) Chapter 1

55   The generals, the institution can select a strategy, lay it all out, but what happens on the battlefield is quite different.

56   [Of science] It give us no answer to our question, what shall we do and how shall we live?
What is Art? 1898

57   If the arrangement of society is bad (as ours is), and a small number of people have power over the majority and oppress it, every victory over Nature will inevitably serve only to increase that power and that oppression. This is what is actually happening.
In Aldous Huxley *Science, Liberty and Peace* 1947

58   I know that most men, including those at ease with problems of the greatest complexity, can seldom accept even the simplest and most obvious truth if it be such as would oblige them to admit the falsity of conclusions which they have delighted in explaining to colleagues, which they have proudly taught to others, and which they have woven, thread by thread, into the fabric of their lives.
J Ford in *Chaotic Dynamics and Fractals* ed M F Barnslay and S G Demko 1985 (New York: Academic Press)

Rudolf Tomaschek 1895–

59   Modern physics is an instrument of Jewry for the destruction of Nordic science.... True physics is the creation of the German spirit.
In W L Shirer *The Rise and Fall of the Third Reich* (London: Secker & Warburg) Chapter 8

Tong Ts'ung-chu 2nd Century BC

60   The fact of harmony between Heaven and Earth and Man does not come from a physical union, from a direct action. It comes from a tuning on the same note producing vibrations in unison.
In B N Parlett *The Symmetric Eigenvalue Problem* 1980 (New York: Prentice-Hall)

Stephen Toulmin 1922–

61   No doubt, a scientist isn't necessarily penalized for being a complex, versatile, eccentric individual with lots of extra-scientific interests. But it certainly doesn't help him a bit.
Civilization and Science in Conflict or Collaboration 1972 (Amsterdam: Elsevier)

Arnold Toynbee 1889–1975

62   [We are in] the first age since the dawn of civilisation in which people have dared to think it practicable to make the benefits of civilisation available to the whole human race.

Thomas Traherne ca 1637–1674

63   He that knows the secrets of nature with Albertus Magnus, or the motions of the heavens with Galileo, or the cosmography of the moon with Hevelius, or the body of man with Galen, or the nature of diseases with Hippocrates, or the harmonies in melody with Orpheus, or of poetry with Homer, or of grammar with Lily, or of whatever else with the greatest artist; he is nothing if he knows them merely for talk or idle speculation, or transient and external use. But he that knows them for value, and knows them his own, shall profit infinitely.

Centuries of Meditations 1908, No 341

Lionel Trilling 1905–1975

64   [Somewhere in the mind] there is a hard, irreducible, stubborn core of biological urgency, and biological necessity, and biological **reason**, that culture cannot reach and that reserves the right, which sooner or later it will exercise, to judge the culture and resist and revise it.

Beyond Culture: Essays on Literature and Learning 1955 (New York: Viking Press)

Lev Davidovich [Bronstein] Trotskii 1879–1940

65   The phenomena of radio-activity lead us straight to the problem of releasing the inner energy of the atom.... The greatest task of contemporary physics is to extract from the atom its latent energy—to tear open a plug so that energy should well up with all its might. Then it will become possible to replace coal and petrol by atomic energy which will become our basic fuel and motive power. This is by no means a hopeless task, and what vistas its solution will open up...scientific and technological thought is approaching the point of a great upheaval; and so the social revolution of our time coincides with a revolution in man's inquiry into the nature of matter and in his mastery of matter.

Speech 1 March 1926. *Sochineniya* XXI, 1927 (Moscow–Leningrad)

66   From the field of chemistry there is no direct and immediate exit to social perspectives.... An objective method of social cognition is necessary. Marxism is that method. When any Marxist tried to convert Marx's theory into a universal skeleton key and flitted through other fields of knowledge, Vladimir Il'ich would rebuke him with the expressive little phrase, 'Communist conceit'. This would signify in particular: Communism does not replace chemistry. But the converse is also true. The attempt to step across Marxism, on the pretext that chemistry (or natural science in general) must solve all problems, is a peculiar chemical conceit, which is theoretically no less erroneous and practically no more likeable than Communist conceit.

D I Mendeleev i Marksizm in *Sochineniya* XXI, 1927 (Moscow–Leningrad)

67   Old age is the most unexpected of all things that can happen to a man.

Diary, 1935. In *Nature* 26 August 1982 **298**

68   Darwin's description of the way in which the pattern on the peacock's feathers formed itself naturally, banished forever the idea of a Supreme Being from my mind.

The Prophet Armed

Harry S Truman 1884–1972

69 We have spent $2,000,000,000 on the greatest scientific gamble in history, and [have] won.
[The atomic bombing of Hiroshima, 6 August 1945]
[The initial 'S' in Harry S Truman does not stand for anything]

Konstantin Eduardovich Tsiolkovskii 1857–1935

70 The Earth is the cradle of mankind. But one does not live in the cradle forever.

Ivan Sergeievich Turgenev 1818–1883

71 Nature is not a temple but a workshop in which man is the labourer.

72 Whatever a man prays for, he prays for a miracle. Every prayer reduces itself to this: 'Great God, grant that twice two be not four'.
Prayer

Turkish Proverb

73 If Allah gives you prosperity, He will give you the brains to go with it.

Mark Twain [Samuel Langhorne Clemens] 1835–1910

74 What a good thing Adam had—when he said a thing he knew nobody had said it before.

75 When I was a boy of 14 my father was so ignorant I could hardly stand to have the old man around. But when I got to be 21, I was astonished at how much he had learnt in 7 years.

76 Scientists have odious manners, except when you prop up their theory; then you can borrow money of them.
The Bee in *What is Man and Other Essays*

77 First get your facts; and then you can distort them at your leisure.

78 There is something fascinating about science. One gets such wholesale returns of conjecture out of such a trifling investment of fact.
Life on the Mississipi

79 It is not what I don't understand in the Bible that worries me—its what I do understand.

80 The researches of many commentators have already thrown much darkness on this subject, and it is probable that, if they continue, we shall soon know nothing at all about it.
The Sciences September–October 1989

Henry Twells 1823–1900

81 When as a child I laughed and wept,
    Time crept.
When as a youth I waxed more bold,
    Time *strolled.*
When I became a full-grown man,
    Time RAN.
When older still I daily grew,

Time FLEW.
Soon I shall find, in passing on,
    Time *gone.*
O Christ! wilt Thou haved saved me then?
    Amen.

[Poem fixed to the front of the clock-case in the North Transept of Chester Cathedral]
Time's Paces in *Newsletter of the Friends of Chester Cathedral* Christmas 1972

John Tyndall 1820–1893

82   I cannot help thinking while I dwell upon them that this discovery of
magnet-electricity [induction] is the greatest experimental result ever obtained by an investigator [Faraday].

Fyodor Ivanovich Tyuchev 1803–1873

83   Nature is not what you might imagine it to be.
It is not blind, it is not a mindless face.
It has a soul, in it is freedom,
In it is love, it has a tongue.

United States Air Force

1   [The United States Air Force ROTC Manual *Fundamentals of Aerospace
Weapons Systems* defines a MILITARY TARGET as] Any person, thing,
idea, entity or location selected for destruction, inactivation, or rendering
non-usable with weapons which will reduce or destroy the will or ability of
the enemy to resist.
[Rapoport draws attention to the mentality which attacks ideas with bombs]
In K Von Clausewitz *On War* ed A Rapoport, 1967 (London: Routledge & Kegan Paul)

US President's Science Advisory Committee

2   In science the excellent is not just better than the ordinary; it is almost all
that matters. It is therefore fundamental that this country should energet-
ically sustain and strongly reinforce first-rate work where it now exists.
Scientific Progress, the Universities and the Federal Government The White House,
Washington, DC, 15 November 1960

Jakob Johann von Uexküll 1854–1944

3   When a dog runs, the dog is moving his legs; when a sea urchin runs, the
legs are moving the sea-urchin.
In Konrad Lorenz *Aspects of Form* ed L L Whyte, 1968 (Indiana)

Mirza Mahommed Ben Sha Rok Ulugh-Beg 1394–1449

4   It is the duty of every true Muslim, man and woman, to strive after knowl-
edge.
[Quoting the Hadith]
Inscribed on the gate of the Ulugh-Beg Medresseh in Bukhara, 1417

Miguel de Unamuno 1864–1937

5   Science is a cemetery of dead ideas.
The Tragic Sense of Life transl P Smith of *Del Sentimiento Trágico de la Vida*, 1953
(London: Routledge & Kegan Paul)

6   *Si un hombre nunca se contradice, será porque nunca dice nada.*
If a man never contradicts himself, it must be because he never says any-
thing.
In E Schrödinger *What is Life?* 1944 (Cambridge: Cambridge University Press)

7   The supreme triumph of reason is to cast doubt on its own validity.

John Updike 1932–

8   Neutrinos, they are very small.
They have no charge and have no mass
And do not interact at all.
The earth is just a silly ball
To them, through which they simply pass,
Like dustmaids down a drafty hall
Or photons through a sheet of glass.
They snub the most exquisite gas,
Ignore the most substantial wall,
Cold shoulder steel and sounding brass,
Insult the stallion in his stall,
And, scorning barriers of class,

Infiltrate you and me. Like tall
And painless guillotines, they fall
Down through our heads into the grass.
At night, they enter at Nepal
And pierce the lover and his lass
From underneath the bed—you call
It wonderful; I call it crass.
Cosmic Gall in *Telegraph Poles and Other Poems* (London: Deutsch)

9   The **Polymers**, those giant Molecules,
Like Starch and Polyoxymethylene,
Flesh out, as protein serfs and plastic fools,
The Kingdom with Life's Stuff.
Our time has seen the synthesis of Polyisoprene
And many cross-linked Helices unknown
To Robert Hooke; but each primordial Bean
Knew Cellulose by heart: **Nature** alone of Collagen
And Apatite compounded Bone.
Dance of the Solids in *Midpoint and other Poems* 1968 (New York: Knopf)

10   Textbooks and Heaven only are ideal
Solidity is an imperfect state.
Within the cracked and dislocated Real
Non-stoichiometric crystals dominate.
Stray Atoms sully and precipitate;
Strange holes, **excitons**, wander loose; because
Of Dangling bonds, a chemical Substrate
Corrodes and catalyzes—surface Flaws
Help epitaxial Growth to fix absorptive claws.
Dance of the Solids in *Midpoint and other Poems* 1968 (New York: Knopf)

James Ussher [Archbishop of Armagh] 1581–1656

11   The world was created on 22nd October, 4004 BC at 6 o'clock in the evening.
Chronologia Sacra Oxford, 1660, p 45. *Annals of the World* Oxford, 1658

Paul Valéry 1871–1945

1   'Science' means simply the aggregate of the recipes that are always successful. All the rest is literature.
Analects vol 14 of Collected Works ed J Matthews, 1970 (London: Routledge & Kegan Paul)

2   Having precise ideas often leads to a man doing nothing.

3   If a man's imagination is stimulated by artificial and arbitrary rules, he is a poet; if it is stifled by such limitations, whatever kind of writer he may be, a poet he is not.

4   *L'histoire est la science des choses qui ne se répétent pas.*
History is the science of things which are not repeated.
Variété IV

5   *L'homme c'est l'homme qu'à sa surface. Lève la peau, dissèque, ici commencent les machines.*
Man is only man at the surface. Remove his skin, dissect, and immediately you come to the machinery.
Oeuvres vol II, 1960 (Paris: Gallimard)

6   One had to be a Newton to notice that the moon is falling, when everyone sees that it doesn't fall.
Analects vol 14 of *Collected Works* ed J Matthews, 1970 (London: Routledge & Kegan Paul)

7   *Il y a* science *des choses simples, et art des choses compliquées.* Science, *quand les variables sont énumerables et leur nombre petit, leur combinaisons nettes et distinctes. On tend vers l'état de science, on le désire. L'artiste se fait des recettes. L'intérêt de la science gît dans* l'art *de faire la science.*
There is a *science* of simple things, an *art* of complicated ones. Science is feasible when the variables are few and can be enumerated; when their combinations are distinct and clear. We are tending toward the condition of science and aspiring to do it. The artist works out his own formulas; the interest of science lies in the *art* of making science.
Analects vol 14 of *Collected Works* ed J Matthews, 1970 (London: Routledge & Kegan Paul)

8   *Tout homme contient une femme. Mais jamais sultane mieux cachée que celle-ci.*
Inside each man there is a woman. But never was even an Arab lady so well hidden as she.
Oeuvres de Paul Valèry I (Paris: Pleiade)

9   Something we see quite clearly, and which nonetheless is very difficult to express, is always worth the trouble of trying to put into words.

Giorgio Vasari 1511–1574

10   He might have been a scientist if he had not been so versatile.
[Of Leonardo da Vinci]
Lives of the Artists ed Burroughs, 1960

The Vatican Council

11   If any one shall not be ashamed to assert that, except for matter, nothing exists; let him be anathema.
Session 3, Canon 2, 24 April 1870

Nikolai Ivanovich Vavilov 1887–1943

12   We shall go to the pyre, we shall burn, but we shall not renounce our convictions.

[The geneticist arrested 6 August 1940; sentenced to death 9 July 1941; elected Foreign Member of the Royal Society 1942; died 26 January 1943]

In Zh A Medvedev *The Rise and Fall of T D Lysenko* 1969 (New York: Columbia University Press)

Sergei Ivanovich Vavilov 1891–1951

13   How have the thematics of scientific research at different times and places been determined and how are they determined? It is only today that we have begun to study this most important problem of the history of science and it is only the Marxists who are doing it.

Marxism and Modern Thought 1935 (London: Routledge)

Thorstein Veblen 1857–1929

14   The outcome of any serious research can only be to make two questions grow where only one grew before.

The Place of Science in Mordern Civilization and Other Essays 1919 (New York: Viking Press)

Paul Verlaine 1844–1896

15   *Fréres, lâchez la science gourmande*
      qui veut voler sur les ceps defendus
      La fruit sanglant qu'il ne faut pas connaître.
      Brothers, touch not gluttonous science
      That off the forbidden vines seeks to steal
      The bloody fruit we must not know.

In Jacques Monod *Inaugural Lecture* 1067 (San Diego, OA. The Salk Institute)
Sagesse XI, 28 (1889)

Vladimir Ivanovich Vernadskii 1863–1945

16   Exact observation of reality shows, that spatial relations—the phenomena of symmetry—lie at the basis of all the phenomena we have studied.

Jules Verne 1828–1905

17   As for the Yankees, they have no other ambition than to take possession of this new continent of the sky [the Moon], and to plant upon the summit of its highest elevation the star-spangled banner of the United States.
      1865

Andreas Vesalius 1514–1564

18   It was when the more fashionable doctors in Italy, in imitation of the old Romans, despising the work of the hand, began to delegate to slaves the manual attentions they deemed necessary for their patients...that the art of medicine went to ruin.

De Humani Corporis Fabrica, in B Farrington *Science in Antiquity* 1936 (Cambridge: Cambridge University Press)

Madame Marie Vichy-Deffand [Marquise du Deffand] 1697–1780

19   *Il n'y a que le premier pas qui coûte.*
It is only the first step which takes the effort.
[Referring to the legend of Saint Denis who walked from his place of execution carrying his head]
Lettre à d'Alembert 1763

[Sir] Geoffrey Vickers 1894–

20   ...the historical causes which produced the Western individual and turned him into the Western individualist. I will not elaborate them here. I would only insist that we should not mistake for laws of God or nature the cultural values of the world's most unstable systems.
Freedom in a Rocking Boat 1970 (London: Penguin)

Vincent de Vigneaud 1901–1978

21   [Canadian biochemist]
Nothing holds up the progress of science so much as the right idea at the wrong time.
In R V Jones *Most Secret War* 1978 (London: Hamish Hamilton)

Eugène-Emanuel Viollet-Le-Duc 1814–1879

22   Symmetry—an unhappy idea for which, in our homes, we sacrifice our comfort, occasionally our commonsense and always a lot of money.
Discourses on Architecture vol 2, transl B Bucknall 1959 (New York: Grove Press)

Rudolf Virchow 1821–1902

23   In my journal, anyone can make a fool of himself.
Archiv für pathologische Anatomie und Physiologie (Zeitschrift für Ethnologie)

24   Pathology is the science of disease [in all organisms] from cells to societies.
Archiv für pathologische Anatomie und Physiologie Introduction

25   The body is a cell state in which every cell is a citizen. Disease is merely the conflict of the citizens of the state brought about by the action of external forces.
Die Cellularpathologie 1858

Virgil [Virgilius Maro] 70–19 BC

26   *Exudent alii spirantia mollius aera*
Credo equidem vivos ducent de marmore vultus
Orabunt causas melius, caelique meatus
Describent radio, et surgentia sidera dicent.
Tu regere imperio populos, Romane, memento,
Hae tibi erunt artes, pacisque imponere morem
Parcere subjectis et debellare superbos.
Others, I know it well, breathing bronze shall trace
And from the deathlike marble call up the living face:
Shall plead with eloquence not thine, shall map and rule the skies,
And with the voice of science shall tell when stars shall set and rise.
'Tis thine O Rome to rule: this mission ne'er forgo
Thine arts, thy science this, to dictate to the foe
To spare who yields submission and bring the haughty low.
Aeneid VI, 851

27   *Felix qui potuit rerum cognoscere causas.*
      Happy is he who gets to know the reasons for things.
      [Motto of Churchill College, Cambridge—local translation: It's great to know what
      makes things tick]
      Georgics II, 490

Marcus Pollio Vitruvius 1st Century BC

28   Water is much more wholesome from earthenware pipes than from lead.
      For it seems to be made injurious by lead, because white lead is produced
      by it; and this is said to be harmful to the human body.
      De Architectura VIII

29   Cracks make...bricks weak.
      Nature 10 February 1983 **301**

30   *Aristippus philosophus Socraticus, naufragio cum eiectus ad Rhodien-*
      sum litus animadvertisset geometrica schemata descripta, exclamavisse ad
      comites ita dicitur; Bene speremus, hominum enim vestiga video.
      The philosopher Aristippus, follower of Socrates, was stranded on the coast
      at Rhodes, and, observing geometrical diagrams drawn upon the sand, he
      is said to have shouted to his companions: 'There are good hopes for us;
      for I see human footsteps!'

Karl Vogt 1817–1895

31   The brain secretes thought as the stomach secretes gastric juice, the liver
      bile, and the kidneys urine.
      Köhlerglaube und Wissenschaft

Tyler Volk

32   How to separate dice from design—that is a major question facing evolu-
      tionary biology today.
      The Sciences May/June 1990

Voltaire [François Marie Arouet] 1694–1778

33   *Vous avez confirmé dans ces lieux pleins d'ennui*
      Ce que Newton connut sans sortir de chez lui.
      You have confirmed in these tedious places
      What Newton found out without leaving his room.
      [On the expedition of Maupertuis to Lapland to confirm the flattening of the Earth at
      the poles]
      4eme Discourse en vers sur l'homme 1737

34   [Invited a second time to an orgy] 'Ah no, my good friends, once a philoso-
      pher, twice a pervert'.

35   There is an astonishing imagination, even in the science of mathematics....
      We repeat, there was far more imagination in the head of Archimedes than
      in that of Homer.
      A Philosophical Dictionary 1881, 3, 40 (Boston)

36   *'Travaillons sans raisonner,' dit Martin; 'c'est le seul moyen de rendre la*
      vie supportable'.
      'Let us work without theorising', said Martin; ' 'tis the only way to make
      life endurable'.
      Candide 1758, XXX. Transl John Butt, 1947 (London: Penguin Classics)

37   *Cunégonde...vit entre les broussailles le docteur-Pangloss qui donnait une leçon de physique expérimentale à la femme de chambre de sa mère, petite brune tres jolie et tres docile.*

One day Cunégonde was walking near the house in a little coppice, called 'the park', when she saw Dr Pangloss behind some bushes giving a lesson in experimental philosophy to her mother's waiting woman, a pretty little brunette who seemed eminently teachable.
Candide 1758, I. Transl John Butt, 1947 (London: Penguin Classics)

38   [Of Newton's post as Master of the Royal Mint] Fluxions and gravitation would have been of no use without a pretty niece.
[Referring to Catherine Barton; wife or mistress of Lord Halifax]

39   *S'étant rétiré, en 1666, à la campagne, prés de Cambridge, un jour qu'il se promenait dans son jardin et qu'il voyait des fruits tomber d'un arbre, il se laissa aller à une méditation profonde sur cette pésanteur, dont tous les philosophes ont cherchés si longtemps en vain, et dans laquel le vulgaire ne soupçonne pas même de mystère...*

Having retreated in 1666 to the countryside near Cambridge, one day he [Newton] was walking in the garden and saw the fruit falling from a tree and he indulged himself in deep meditation on gravity, about which philosophers have so long searched in vain, and in which the common people do not even suspect a mystery....
[It is thus Voltaire who has preserved this story which he had from Newton's niece, Mrs Conduitt]
Lettres Philosophiques, lettre 15, 1734

40   Man can have only a certain number of teeth, hairs and ideas; there comes a time when he necessarily loses his teeth, hair and ideas.
Dictionnaire Philosophique 1764

41   *Un Français qui arrive a Londres trouve les choses bien changées, en philosophie comme tout le reste. Il a laissé le monde plein; il le trouve vide. À Paris, on voit l'univers composé de tourbillons de matière subtile; à Londres on ne voit rien de çela.*

A Frenchman who arrives in London, will find philosophy, like everything else, very much changed there. He had left the world a plenum, and he now finds it a vacuum. At Paris the universe is seen composed of vortices of subtle matter; but nothing like it is seen in London.
Letters Concerning the English Nation 1733. *Lettres Philosophiques*, lettre 14, *On Descartes and Newton* 1727

[Field-Marshal] Helmuth Carl Bernard Von Moltke 1800–1891

42   No plan survives contact with the enemy.

Richard Frederick Voss 1948–

43   It seems obvious that painting, sculpture or drama imitated nature. But what does music imitate? The measurements suggest that music is imitating the characteristic way our world changes in time.
[On 1/*f* noise]
Science and Uncertainty 1985, *IBM/Science Reviews Ltd.*

Alexander Vucinich 1914–

44   Every scientist is an agent of cultural change. He may not be a champion of change; he may even resist it, as scholars of the past resisted the

new truths of historical geology, biological evolution, unitary chemistry, and non-Euclidean geometry. But to the extent that he is a true professional, the scientist is inescapably an agent of change. His tools are the instruments of change—skepticism, the challenge to established authority, criticism, rationality, and individuality.

Science in Russian Culture: A History to 1860 1963 (Stanford, CA: Stanford University Press)

Conrad Hal Waddington 1905–1975

1   Science is the organised attempt of mankind to discover how things work
     as causal systems. The scientific attitude of mind is an interest in such
     questions. It can be contrasted with other attitudes, which have different
     interests; for instance the magical, which attempts to make things work
     not as material systems but as immaterial forces which can be controlled
     by spells; or the religious, which is interested in the world as revealing the
     nature of God.
     The Scientific Attitude 1941 (London: Penguin)

2   DNA plays a role in life rather like that played by the telephone directory
     in the social life of London: you can't do anything much without it, but,
     having it, you need a lot of other things—telephones, wires, and so on—as
     well.
     Review of *The Double Helix* 25 May 1968 *The Sunday Times*

George Wald 1906–

3   We are the products of editing, rather than of authorship.
     In *The Origin of Optical Activity* from the *Annals of the New York Academy of
     Sciences* 1975 **69** 352–68

4   We already know enough to begin to cope with all the major problems that
     are now threatening human life and much of the rest of life on earth. Our
     crisis is not a crisis of information; it is a crisis of decision of policy and
     action.
     Philosophy and Social Action V, (1–2), 1979

5   It would be a poor thing to be an atom in a universe without physicists,
     And physicists are made of atoms. A physicist is an atom's way of knowing
     about atoms.
     Foreword to L J Henderson *The Fitness of the Environment* 1958 (Boston: Beacon)

Alfred Russell Wallace 1823–1913

6   In proportion as physical characteristics become of less importance, mental
     and moral qualities will have an increasing importance to the well-being of
     the race. Capacity for acting in concert, for protection of food and shelter;
     sympathy, which leads all in turn to assist each other; the sense of right,
     which checks depredation upon our fellows…all qualities that from earliest
     appearance must have been for the benefit of each community, and would
     therefore have become objects of natural selection.
     Origin of human races and the antiquity of man in *Journal of the Royal Anthropo-
     logical Society, London* 1864, clviii

7   These checks—war, disease, famine and the like—must, it occurred to me,
     act on animals as well as man. Then I thought of the enormously rapid
     multiplication of animals, causing these checks to be much more effective
     in them than in the case of man; and while pondering vaguely on this fact
     there suddenly flashed upon me the idea of the survival of the fittest—
     that the individuals removed by these checks must be on the whole inferior
     to those that survived. In the two hours that elapsed before my ague fit
     was over, I had thought out almost the whole of the theory: and the same
     evening I sketched the draft of paper, and in the two succeeding evenings
     wrote it out in full, and sent it by the next post to Mr. Darwin.
     In B Willey *Darwin and Butler* 1960 (London: Chatto & Windus)

8   We permit absolute possession of the soil of our country, with no legal rights
of existence on the soil to the vast majority who do not possess it. A great
landholder may legally convert his whole property into a forest or a hunting-
ground, and expel every human being who has hitherto lived upon it. In a
thickly-populated country like England, where every acre has its owner and
its occupier, this is a power of legally destroying his fellow-creatures; and
that such a power should exist, and be exercised by individuals, in however
small a degree, indicates that, as regards true social science, we are still in
a state of barbarism.
The Malay Archipelago [last words] 1869 (London: Macmillan)

Henry A Wallace 1888–1965

9   [US Secretary of Agriculture, 1940]
Science, of course, is not like wheat or automobiles. It cannot be
overproduced.... In fact, the latest knowledge is usually the best. Moreover,
knowledge grows or dies. It cannot live in cold storage. It is perishable and
must be constantly renewed.
Scientific American September 1982

Graham Wallas 1958–1932

10   ...'How can I know what I think till I see what I say?'
The Art of Thought ed May Wallas, 1945 (London: Watts)

John Wallis 1616–1703

11   Whereas Nature does not admit of more than three dimensions...it may
justly seem very improper to talk of a solid...drawn into a fourth, fifth,
sixth, or further dimension.
Algebra 1685

Horace Walpole 1717–1797

12   The natural philosophers in power believe that Dr. Franklin has invented
a machine the size of a toothpick case and materials that would reduce St.
Paul's to a handful of ashes.
1778. In Carl van Doren *Benjamin Franklin*

Izaak Walton 1593–1683

13   Angling may be said to be so like mathematics, that it can never be fully
learnt.
The Compleat Angler, Epistle to the Reader, 1653

Wang Yang Ming [Wang Shou-Jen] 1472–1528

14   To know and not to act is not yet to know.

Mary Helen [Baroness] Warnock 1924–

15   To be a woman intellectual is still fraught with contradictions.
The Observer 3 June 1990

Robert Penn Warren 1905–

16   ...What if angry vectors veer
Round your sleeping head, and form.
There's never need to fear

Violence of the poor world's abstract storm.
Lullaby in *Encounter* May 1957

James Dewey Watson 1928–

17   It is necessary to be slightly underemployed if you want to do something
significant.
In H Judson *The Eighth Day of Creation*

18   No *good* model ever accounted for *all* the facts, since some data was bound
to be misleading if not plain wrong.
In Francis Crick *Some Mad Pursuit* 1988 (London: Weidenfeld & Nicolson)

James Watt 1736–1819

19   James Watt, Who, directing the force of an original genius, Early exercised
in philosophic research to the improvement of THE STEAM ENGINE,
enlarged the resources of his country, increased the power of men, and rose
to an eminent place among the most illustrious followers of science and the
real benefactors of the world.
Epitaph in Westminster Abbey, London

Warren Weaver 1894–

20   The century of biology upon which we are now well embarked is no matter
of trivialities. It is a movement of really heroic dimensions, one of the
great episodes in man's intellectual history. The scientists who are carrying
the movement forward talk in terms of nucleo-proteins, of ultracentrifuges,
of biochemical genetics, of electrophoresis, of the electron microscope, of
molecular morphology, of radioactive isotopes. But do not be fooled into
thinking this is mere gadgetry. This is the dependable way to seek a solution
of the cancer and polio problems, the problem of rheumatism and of the
heart. This is the knowledge on which we must base our solution of the
population and food problems. This is the understanding of life.
Letter to H M H Carson in R B Fosdick *The Story of the Rockefeller Foundation* 1952
(New York: Harper)

G Weber

21   The protein molecule model resulting from the X-ray crystallographic observations is a 'Platonic' protein, well removed in its perfection from the
kicking and screaming 'stochastic' molecule that we must infer must exist
in solution.
Advances in Protein Chemistry 1975 **29** 2

Wei Cheng T'ang Dynasty

22   Hearing both sides brings enlightenment.
Believing only one side brings obscurity.
Sakata Shoichi *Suppl. Prog. Theor. Phys.* 1971 **50** 211. See also: Mao Tse Tung *On
Contradiction* 1937

Simone Weil 1909–1943

23   *La science, aujourd'hui, cherchera une source d'inspiration audessus d'elle
ou périra.*
*La science ne présente que trois intérêts: 1. les applications techniques;
2. jeu d'échecs; 3. chemin vers Dieu. (Le jeu d'échecs est agrémenté de*

concours, prix et medailles.)
Science today must search for a source of inspiration higher than itself or it must perish.
Science offers only three points of interest: 1. technical applications; 2. as a game of chess; 3. as a way to God. (The chess-game is embellished with competitions, prizes and medals.)
La Pesanteur et la Grace 1967 (Paris: Librairie Plon)

Alvin M Weinberg 1915–

24   I would therefore sharpen the criterion of scientific merit by proposing that, other things being equal, that field has the most merit which contributes most heavily to, and illuminates most brightly, its neighbouring scientific disciplines.
Minerva 1963 **2** 159–71

Gerald M Weinberg

25   The average scientist is good for at most one revolution. Even if he has the power to make one change in his category system and carry others along, success will make him a recognised leader, with little to gain from another revolution.
An Introduction to General Systems Thinking 1975 (New York: Wiley)

26   If you can't stand the microwaves, keep out of the kitchen.
An Introduction to General Systems Thinking 1975 (New York: Wiley)

Paul Alfred Weiss 1898–

27   A system...is exactly the opposite of a machine, in which the structure of the product depends crucially on strictly predefined operations of the parts. In the system, the structure of the whole determines the operation of the parts; in the machine, the operation of the parts determines the outcome.
Beyond Reductionism ed A Koestler and V R Smithies, 1968 (London: Hutchinson)

28   *Omnis organisatio ex organisatione.*
Organisation only from organisation.
1940. In D J Haraway *Crystals, Fabrics and Fields* 1976 (New Haven, CT: Yale University Press)

Paul Alfred Weiss 1901–

29   [Of the Rockefeller Institute] We must convince the scientific community that chemical warfare and biological warfare is not a dirty business. It is no worse than other means of killing.
[National Meeting of the American Chemical Society, Cleveland, February 1961. *Symposium on Non-military Defense—Chemical and Biological Defenses in Perspective.*]
John Barden *Bulletin of the Atomic Scientists* February 1961

Victor Frederick Weisskopf 1908–

30   Science has become adult; I am not sure whether scientists have.
Scientists in Search of their Conscience ed A R Michaelis and H Harvey

31   The value of fundamental research does not lie only in the ideas it produces. There is more to it. It affects the whole intellectual life of a nation by determining its way of thinking and the standards by which actions and intellectual production are judged. If science is highly regarded and if the

importance of being concerned with the most up-to-date problems of fundamental research is recognized, then a spiritual climate is created which influences the other activites. An atmosphere of creativity is established which penetrates every cultural frontier. Applied sciences and technology are forced to adjust themselves to the highest intellectual standards which are developed in the basic sciences. This influence works in many ways: some fundamental students go into industry; the techniques which are applied to meet the stringent requirements of fundamental research serve to create new technological methods. The style, the scale, and the level of scientific and technical work are determined in pure research; that is what attracts productive people and what brings scientists to those countries where science is at the highest level. Fundamental research sets the standards of modern scientific thought; it creates the intellectural climate in which our modern civilization flourishes. It pumps the lifeblood of idea and inventiveness not only into the technological laboratories and factories, but into every cultural activity of our time. The case for generous support for pure and fundamental science is as simple as that.

Why pure science? in the *Bulletin of the Atomic Scientists* 1965 **21** 4–8

Chaim Weizmann 1874–1952

32   [To Meyer Weisgal, Director of the Weizmann Institute] Never let the scientists get near the *Shissel* [container, where the money is kept].
In Joseph Wechsberg *A Walk Through the Garden of Science* 1967 (London: Weidenfeld & Nicolson)

Carl Friedrich von Weizsäcker 1912–

33   Those reductionists who try to reduce life to physics usually try to reduce it to primitive physics—not to good physics. Good physics is broad enough to contain life, to encompass life in its description since good physics allows a vast field of possible descriptions. There is no reason why living beings should be compared to primitive machines which don't make use of feedback.
Theoria to Theory 1968, vol 3

Arthur Mellen Wellington 1847–1895

34 Engineering...is the art of doing that well with one dollar, which any bungler can do with two after a fashion.
The Economic Theory of the Location of Railways 6th edn, 1900 (New York: Wiley)

Arthur Wellesley [Duke of] Wellington 1769–1852

35 All the business of war, and indeed all the business of life, is an endeavour to find out what you don't know by what you do; that's what I call guessing what was on the other side of the hill.
[Said of the Duke of Brunswick, invading Revolutionary France, that he turned back because 'he did not know what was on the other side of the hill']

Herbert George Wells 1866–1946

36 ...my epitaph. That, when the time comes, will manifestly have to be: 'I told you so. You *damned* fools'. [The italics are mine.]
Preface to the 1941 edition of *The War in the Air* originally written in 1907

37 I must confess that I believe quite firmly that an inductive knowledge of a great number of things in the future is becoming a human possiblity. So far nothing has been attempted, so far no first-class mind has ever focused itself upon these issues. But suppose the laws of social and political development, for example, were given as many brains, were given as much attention, criticism and discussion as we have given to the laws of chemical composition during the last fifty years—what might we not expect?
The Discovery of the Future Lecture at the Royal Institution, 24 January 1902

38 In England we have come to rely upon a comfortable time-lag of fifty years or a century intervening between the perception that something ought to be done and a serious attempt to do it.
The Work, Wealth and Happiness of Mankind 1932 (London: Heinemann) Chapter 2

39 Queen Victoria was like a great paper-weight that for half a century sat upon men's minds, and when she was removed their ideas began to blow about all over the place haphazardly.
The Time Traveller 1973 (London: Weidenfeld & Nicolson)

40 There comes a moment in the day, when you have written your pages in the morning, attended to your correspondence in the afternoon, and have nothing further to do. Then comes that hour when you are bored; that's the time for sex.
In N Mackenzie and J Mackenzie *The Time Traveller H G Wells* 1973 (New York: Simon & Schuster)

41 No one would have believed in the last years of the nineteenth century that this world was being watched keenly and closely by intelligences greater than man's and yet as mortal as his own; that as men busied themselves about their various concerns, they were scrutinised and studied, perhaps almost as narrowly as a man with a microscope might scrutinise the transient creatures that swarm and multiply in a drop of water.
The War of the Worlds 1897. (Opening sentence)

42 Now you begin to see the object of my investigations into the geometry of four dimensions. Long ago I had a vague inkling of a machine.
The Time Machine

43   Biologists can be just as sensitive to heresy as theologians.
      The International Dictionary of Thoughts (Chicago: J G Ferguson Publishing Co)

44   Statistical thinking will one day be as necessary for efficient citizenship as
      the ability to read and write.
      In Warren Weaver *Scientific American* January 1952

45   History becomes more and more a race between education and catastrophe.
      Federico Mayor, 8 February 1990

John Wesley 1703–1791

46   [The founder of the Methodist sect]
      Electricity is the soul of the universe.

Rebecca West 1892–1983

47   Before a war military science seems a real science, like astronomy; but after
      a war it seems more like astrology.

Hermann Weyl 1885–1955

48   The whole is always more, is capable of a much greater variety of wave
      states, than the combination of its parts.... In this very radical sense,
      quantum physics supports the doctrine that the whole is more than the
      combination of its parts.
      Philosophy of Mathematics and Natural Science 1949 (Princeton, NJ: Princeton Uni-
      versity Press)
      [See *Aristotle*]

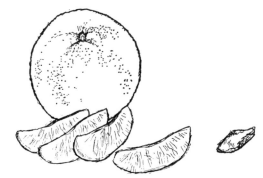

John Archibald Wheeler 1911–

49   There is nothing in the world except empty curved space. Matter, charge,
      electromagnetism and other fields are only manifestations of the curvature
      of space.
      1957. In *New Scientist* 26 September 1974

50   Time is defined so that motion looks simple.
      Gravitation 1973 (Reading: Freeman)

51   The laws of physics must provide a mechanism for the universe to come
      into being.

52 Time is what prevents everything from happening at once.
American Journal of Physics 1978 **46** 323
[But also C J Overbeck: 'Time is that great gift of nature which keeps everything from happening at once.']

53 What are these seven versions of our trilogy of action [for 'seekers of the larger view']:
1. Discover unity.
2. Draw together pieces of science and technology to *create a system*, whether that system is xerography, telegraphy or steam navigation.
3. Find the *economic feasibility* for a new technology by virtue of a wide grasp of the worlds of man and matter.
4. *Reach harmony through intuition*, by meditating on the base of a wide and deep knowledge of the field so as to arrive at a new result.
5. *Build a model*, a simplified representation of the problem at issue, subject to experimental and calculational analysis.
6. *Serve as a science-technologist generalist* who, not once or twice in his life, but many times in a year, and generally in the service of others, extracts the single, simple missing point out of a complicated situation.
7. *Make decisions*, or help others make decisions, *by imaginative interaction with alternative scenarios* calculated as consequent on those decisions.

William Whewell 1794–1866

54 As we read the *Principia* [of Newton] we feel as when we are in an ancient armoury where the weapons are of gigantic size; and as we look at them we marvel what manner of man was he who could use as a weapon what we can scarcely lift as a burden.
In E N da C Andrade *Newton and the Science of his Age. Proceedings of the Royal Society* 6 May 1943

55 We need very much a name to describe a cultivator of science in general. I should incline to a call him a scientist.
[The first use of the word]
The Philosophy of the Inductive Sciences vol I 1840

'Whipplesnaith'

56 The dons also give no trouble.... The younger dons, indeed, are often roof climbers themselves. Out of a bare score whom the writer knows, four are active roof-climbers, and he knows of another four who have each reached the top of King's Chapel, usually reckoned to be the biggest climb in Cambridge. In fact, if you tactfully broach the subject to your supervisor, he may be able to help you considerably.
The Night-Climbers of Cambridge 1937 (London: Chatto & Windus)

James Abbott McNeil Whistler 1834–1903

57 [Answering Oscar Wilde's 'I wish I had said that'] You will, Oscar, you will.
In L C Ingleby *Oscar Wilde*

58 If silicon had been a gas, I should have been a Major General.
[Explaining his failure at the chemistry paper in the West Point entrance examination]
TIBS July 1978

Alfred North Whitehead 1861–1947

59   The aims of scientific thought are to see the general in the particular and the eternal in the transitory.

60   A crystal lacks rhythm from excess of pattern, while a fog is unrhythmic in that it exhibits a patternless confusion of detail.
The Principles of Natural Knowledge 1925 (Cambridge)

61   It is a profoundly erroneous truism, repeated by all copy books and by eminent people when they are making speeches, that we should cultivate the habit of thinking of what we are doing. The precise opposite is the case. Civilization advances by extending the number of important operations which we can perform without thinking about them.
Introduction to Mathematics 1911 (New York: Williams & Norgate). In *Scientific American* September 1966

62   It is a safe rule to apply that, when a mathematical or philosophical author writes with a misty profoundity, he is talking nonsense.
An Introduction to Mathematics 1911 (New York: Williams & Norgate)

63   The possibilities of modern technology were first in practice realised in England by the energy of a prosperous middle class. Accordingly, the industrial revolution started there. But the Germans explicitly realised the methods by which the deeper veins in the mine of science could be reached. In their technological schools and universities progress did not have to wait for the occasional genius or the occasional lucky thought. Their feats of scholarship during the nineteenth century were the admiration of the world. This discipline of knowledge applies beyond technology to pure science, and beyond science to general scholarship. It represents the change from amateurs to professionals.
Science and the Modern World 1926 (London: Cambridge University Press)

64   Science is taking on a new aspect which is neither purely physical nor purely biological. It is becoming the study of organisms. Biology is the study of the larger organisms; whereas physics is the study of the smaller organisms.
Science and the Modern World 1926 (London: Cambridge University Press)

65   A science which hesitates to forget its founders is lost.

66   [In reply to complaints from the Bureau of Inland Revenue] I never write letters, If I wrote letters how could I possibly do my work.
In Alistair Cooke *The Americans* 1979 (London: Bodley Head). Letter of 10 February 1972

67   Youth is imaginative, and if the imagination can be strengthened by discipline, this energy of imagination can in great measure be preserved through life. The tragedy of the world is that those who are imaginative have but slight experience, and those who are experienced have feeble imagination. Fools act on imagination without knowledge, pedants act on knowledge without imagination. The task of a university is to weld together imagination and experience.
Nature 18 November 1982 **300** 296

68   Knowledge does not keep any better than fish.
The Aims of Education section III, Chapter vii

Walt Whitman ca 1819–1892

69   Do I contradict myself? Very well then I contradict myself (I am large, I contain multitudes).
Song of Myself 1938 (London: Nonesuch Press) 5

70   I need no assurances...
I do not doubt that temporary affairs keep on and on millions of years.
I do not doubt that interiors have their interiors and exteriors have their exteriors...
Assurances in *Nonesuch Edition of Collected Poems* 1938 (London: Nonesuch Press)

71   When I heard the learn'd astronomer,
When the proofs, the figures, were ranged in columns before me,
When I was show the charts and diagrams, to add, divide, and measure them,
When I sitting heard the astronomer where he lectured with much applause in the lecture-room,
How soon unaccountable I became tired and sick,
Till rising and gliding out I wander'd off by myself,
In the mystical moist night-air, and from time to time,
Look'd up in perfect silence at the stars.
[1865–1867]
When I heard the learn'd astronomer in *Nonesuch Edition of Collected Poems* 1938 (London: Nonesuch Press)

Benjamin Lee Whorf 1897–1934

72   ...an explicit scientific world view may arise by a higher specialization of the same basic grammatical patterns that fathered the naive and implicit view. Thus the world view of modern science arises by higher specialization of the basic.
Language, Thought and Reality 1956 (Cambridge, MA: The MIT Press)

Lancelot Law Whyte 1896–1972

73   Both science and art have to do with ordered complexity.
The Griffin 1957 **6** No 10

Sheila Evans Widnall 1938–

74   [Chairman of the Faculty committee on undergraduate admissions, MIT]
Once experienced, the expansion of personal intellectual power made available by the computer is not easily given up.
Science 12 August 1983 **221** 607

Norbert Wiener 1894–1964

75   A painter like Picasso, who runs through many periods and phases, ends up by saying all those things which are on the tips of the tongues of the age to say, and finally sterilises the originality of his contemporaries and juniors.
The Human Use of Human Beings 1950 (London: Sphere Books)

76   We are raising a generation of young men who will not look at any scientific project which does not have millions of dollars invested in it.... We are for the first time finding a scientific career well paid and attractive to a large

number of our best young go-getters. The trouble is that scientific work
of the first quality is seldom done by the go-getters, and that the dilution
of the intellectual milieu makes it progressively harder for the individual
worker with any ideas to get a hearing.... The degradation of the position
of the scientist as an independent worker and thinker to that of a morally
irresponsible stooge in a science-factory has proceeded even more rapidly
and devastatingly than I had expected.
Bulletin of the Atomic Scientists 4 November 1948, pp 338–9

Jerome Bert Wiesner 1915–

77   Some problems are just too complicated for rational logical solutions. They
admit of insights, not answers.
In D Lang Profiles: A Scientist's Advice II. In New Yorker 26 January 1963

Eugene Paul Wigner 1902–

78   The simplicities of natural laws arise through the complexities of the languages we use for their expression.
Communications on Pure and Applied Mathematics 1959 **13** 1

79   The language of mathematics reveals itself [to be] unreasonably effective
in the natural sciences...a wonderful gift which we neither understand nor
deserve. We should be grateful for it and hope that it will remain valid
in future research and that it will extend, for better or for worse, to our
pleasure even though perhaps also to our bafflement, to wide branches of
learning.
1960

Richard Wilbur 1921–

80   [Nobel Laureate]
Kick at the rock, Sam Johnson, break your bones;
But cloudy, cloudy is the stuff of stones.
We milk the cow of the world, and as we do
We whisper in her ear, 'You are not true.'
Epistemology

Oscar Wilde 1854–1900

81   Experience is the name every one gives to their mistakes.
Lady Windermere's Fan in *Complete Works of Oscar Wilde* 1966 (London: Collins)

82   Religions die when they are proved to be true. Science is the record of dead
religions.
Phrases and Philosophies for the Use of the Young 1894. *Complete Works of Oscar
Wilde* 1966 (London: Collins)

83   In England, at any rate, education produces no effect whatsoever. If it did,
it would prove a serious danger to the upper classes, and probably lead to
acts of violence in Grosvenor square.
[Which it did, in the riots of 1968 in front of the American Embassy]
The Importance of Being Earnest

84   Sonnet to Liberty.
Not that I love thy children,...
...and yet, and yet,

These Christs that die upon the barricades,
God knows that I am with them, in some ways.
Complete Works of Oscar Wilde 1966 (London: Collins)

85 Heredity is the last of the fates, and the most terrible.
Anthony Smith *The Human Pedigree* 1975 (London: Allen & Unwin)

86 [On the Niagara Falls] It would be more impressive if it flowed the other way.
Science 19 August 1977 **197** 742

87 Nature has good intentions, of course, but, as Aristotle once said, she cannot carry them out.
The Decay of Lying

Helmuth Wilhelm 1905–

88 Change, that is the only thing in the Universe which is Unchanging.
[Epitomizing the *I Ching*, 8th Century BC and quoting Heraclitus]
Der Zeitbegriff im Buch der Wandlungen in *Eranos Jahrbuch* 1951 **20** 321

William Henry [Duke of Gloucester] 1743–1805

89 Another damned, thick, square book. Always scribble, scribble, scribble. Eh: Mr Gibbon?
Quoted in a Note to Boswell's *Life of Johnson* vol ii

Shirley Williams 1930–

90 [Minister for Education and Science]
[For scientists] the party is over.
1971. Quoted by A Cottrell, Royal Institution Lecture, 10 November 1978

[Mr Justice] Wills

91 It is wise in any state to encourage letters and the painful researches of learned men. The easiest and most equal way of doing it is by securing to them the property of their own work. A winter's fame will not be the less that he had bread without being under the necessity of prostituting his pen to flattery to get it. He who engages in laborious work (such for instance as Johnson's Dictionary) which may employ his whole life, will do it with more spirit if besides his own glory he thinks it may be a provision for his family.
Miller vs. Taylor. In *ASLIB Proc.* 1979 **31** No 2 52

[Lord] John [of Selmeston] Wilmot 1895–1964

92 What I like about scientists is that they are a team, so that one need not know their names.
[Minister of Supply 1945–1947]
The Prof in two Worlds 1961 (London: Collins)

Charles G Wilson

93 [US Secretary of Defense]
Basic research is when you don't know what you are doing.
Nature 1976 **264** 100

Edward Osborne Wilson 1929–

94　Marxism is sociobiology without biology.... Although Marxism was formulated as the enemy of ignorance and superstition, to the extent that it has become dogmatic it has faltered in that commitment and is now mortally threatened by the discoveries of human sociobiology.
On Human Nature 1978 (Cambridge, MA: Harvard University Press)

95　Let me use this famous painting by Paul Delaroche of Cardinal Mazarin on his deathbed to illustrate the current stance of what might be called conventional sociobiology.... The great prelate, advisor to Louis the Fourteenth, is dying in 1661, ostensibly without produce, without producing any replicates of Mazarin DNA. But is he? In this painting we find him surrounded by his nieces and their husbands, who have been raised to high places in the French court and who are prospering mightily and having nephews and grand-nephews and grand-nieces for the cardinal.
R Wright *Three Scientists and Their Gods* 1988 (New York: Times Books)

[Sir] Harold Wilson 1916–

96　If there was one word I could use to identify modern socialism it was 'science'.
[Speech at Daresbury, 17 June 1967]
[But the 'white-hot technological revolution' never took place and, like Christianity, science was never really tried]
The Relevance of British Socialism 1964 (London: Weidenfeld & Nicolson)
See A Cottrell *Physics Bulletin* March 1976 p102

97　The Britain that is going to be forged in the white heat of this revolution will be no place for restrictive practices or out of date methods, on either side of industry.
Labour Party Conference, Scarborough, 30 September to 4 October 1963. In B Lovell
P M S Blackett—a Biographical Memoir, Royal Society, London, 1976

Wallace Wilson 20th Century

98　He prayeth best who loveth best
All creatures great and small.
The Streptococcus is the test
I love him least of all.
[In parody of Coleridge's *Ancient Mariner*]

Ludwig Wittgenstein 1889–1951

99　In order to draw a limit to thinking, we should have to think both sides of this limit.
Tractatus Logico–Philosophicus 1961 (London: Routledge & Kegan Paul)

100　We could present spatially an atomic fact which contradicted the laws of physics, but not one which contradicted the laws of geometry.
Tractatus Logico–Philosophicus 1961 (London: Routledge & Kegan Paul)

101　We feel that even if all possible scientific questions be answered, the problems of life have still not been touched at all. Of course there is then no question left, and just this is the answer. The solution of the problem of life is seen in the vanishing of this problem.
Tractatus Logico–Philosophicus 1961 (London: Routledge & Kegan Paul)

102 Whereof one cannot speak, thereof one must be silent.
Tractatus Logico-Philosophicus 1961 (London: Routledge & Kegan Paul)

103 *Handeln von Netz, nicht von dem, was das Netz beschreibt.*
Deal with the network, not with what the network represents.
[The totality of existent atomic facts is the world]
Phil. Trans. RS 1979/1980 **A295** 554

Friedrich Wöhler 1800–1882

104 Organic chemistry just now is enough to drive one mad. It gives one the impression of a primeval, tropical forest full of the most remarkable things, a monstrous and boundless thicket, with no way of escape, into which one may well dread to enter.
Letter to Berzelius 28 January 1835

Roberta Wohlstetter

105 To understand the fact of surprise it is necessary to examine the characteristics of the noise as well as the signals that after the event are clearly seen to herald the attack.
Pearl Harbour: Warning and Decision 1962 (Stanford, CA: Stanford University Press)

[Cardinal] Thomas Wolsey ca 1475–1530

106 This new invention of printing has produced various effects of which Your Holiness cannot be ignorant. If it has restored books and learning, it has also been the occasion of those sects and schisms which daily appear. Men begin to call in question the present faith and tenets of the Church; the laity read the scriptures and pray in their vulgar tongue. Were this suffered the common people might come to believe that there was not so much use of the clergy. If men were persuaded that they could make their own way to God, and in their ordinary language as well as Latin, the authority of the Mass would fall, which would be prejudicious to our ecclesiastical orders. The mysteries of religion must be kept in the hands of the priests.

John Woodall

107 [16th Century surgeon]
It is no small thing to dismember the image of God.
In Joan Cassell *Anthropology Today* 1986 **2** No 2 13–15

Joseph Henry Woodger 1894–1952

108 An organism consists of chemical atoms plus ordering relations.
In L L Whyte *Internal Factors in Evolution* 1965 (London: Tavistock)

Pelham Grenville Woodhouse 1881–1975

109 [Jeeves to Bertie Wooster:] 'You would not enjoy Nietzsche, Sir. He is fundamentally unsound.'

Frederic Wood-Jones 1879–1954

110 Whoever wins to a great scientific truth will find a poet before him in the quest.
Medical Journal of Australia 29 August 1931

[Sir] Richard van der Riet Woolley 1906–1986

111   Space-travel is utter bilge.
[The Astronomer-Royal, 1956–1971. Quoted ca 1956, see *Obit. Mem. FRS*]

[Baroness] Barbara Frances Wootton 1897–

112   Of the two great superstitions of the Western world—Christianity and
Marxism—Christianity has, of course, had a much longer period in which
to sterilize fine intelligences and divert the most powerful instrument that
we know—the human mind—from fruitful use in the service of the species
which possess it.
Testament for Social Science 1950 Chapter IV, 2

Elizabeth Wordsworth 1840–1932

113   If all the good people were clever;
And all the clever people were good,
The world would be nicer than ever
We thought that it possibly could.
Good and Clever

William Wordsworth 1770–1850

114   ...Where the statue stood
Of Newton with his prism and silent face,
The marble index of a mind for ever
Voyaging through strange seas of Thought, alone.
[Newton's Statue at Trinity College, Cambridge]
The Prelude 1850, Book III, written 1779–1805

115   Lost in a gloom of uninspired research.
The Excursion Book 4

116   Man now presides
In power, where once he trembled in his weakness;
Science advances with gigantic strides;
But are we aught enriched in love and meekness?
To the Planet Venus 1838

117   Physician art thou—one, all eyes,
Philosopher—a fingering slave,
One that would peep and botanize
Upon his mother's grave.
A Poet's Epitaph

118   Poetry is the breath and finer spirit of all knowledge; it is the impassioned
expression which is in the countenance of all Science...shall be ready to
put on, as it were, a form of flesh and blood, the poet will lend his divine
spirit to aid the transfiguration, and will welcome the Being thus produced
as a dear and genuine inmate of the household of man.

119   To the solid ground of nature trusts the Mind that builds for aye.
Quotation appearing on the title page of *Nature* until 1963

120   Yet we may not entirely overlook
The pleasure gathered from the rudiments
Of geometric science...
The Prelude 1850, Book VI, lines 115–7

121 [Mathematics]...is an independent world,
Created out of pure intelligence.

Frank Lloyd Wright 1869–1959

122 An expert is a man who has stopped thinking—he knows.
The Computer Age ed M L Dertouzos and J Moses 1979 (Cambridge, MA: MIT Press)

John Wycliffe ca 1320–1384

123 God forceth not a man to believe that which he cannot understand.
[Translator of the Bible into English]

Xenophon ca 444–ca 354 BC

1   What are called the mechanical arts carry a social stigma and are rightly dishonoured in our cities. For these arts damage the bodies of those who work at them or who act as overseers, by compelling them to a sedentary life and, in some cases, to spend the whole day by the fire. This physical degeneration results also in deterioration of the soul. Furthermore, the workers in these trades simply have not got the time to perform the offices of friendship or citizenship. Consequently they are looked on as bad friends and bad patriots, and in some cities, especially the warlike ones, it is not legal for a citizen to ply a mechanical trade.
Oeconomicus
In B Farrington *Greek Science* 1963 (London: Penguin)

Yang Hsiung 51 BC–18 AD

1   Someone asked whether a sage could make divination. [Yang Hsiung] replied that a sage could certainly make divination about Heaven and Earth. If that is so, continued the questioner, what is the difference between the sage and the astrologer (*shih*)? [Yang Hsiung] replied, 'The astrologer foretells what the effects of heavenly phenomena will be on man; the sage foretells what the effects of man's actions will be on the heavens'.
    Fa Yen (Model Discourses) ca 5 AD. Transl J Needham *Science and Civilization in China* 1956 (London: Cambridge University Press)

Yamagiwa Katsusaburo 1863–1930

2   Cancer has been created!
    With pride, I advance a few steps.
    [17-syllable haiku written on producing cancer by painting the ears of rabbits with coal-tar]
    1915. *Profiles of Japanese Science and Scientists* ed H Yukawa 1970 (Tokyo: Kodansha)

William Butler Yeats 1865–1939

3   The intellect of man is forced to choose
    Perfection of the life, or of the work.
    The Choice in *The Collected Poems of W B Yeats* 1933 (London: Macmillan)

4   Locke sank into a swoon;
    The Garden died;
    God took the spinning-jenny
    Out of his side.
    Fragments in *The Colleted Poems of W B Yeats* 1933 (London: Macmillan)

5   Science is the religion of the suburbs.
    In *The Spectator* 5 July 1969

Robert M Yerkes 1876–1956

6   One chimpanzee is not a chimpanzee at all.
    Chimpanzees, a Laboratory Colony 1945 (New Haven, CN: Yale University Press)

Edward Young 1683–1765

7   Love finds admission, where proud science fails.
    Night Thoughts 1742–1745

8   This gorgeous apparatus
    This display
    This ostentation of creative power
    This theatre
    What eye can take it in...?
    The Complaint, or Night Thoughts, Night Ninth 1744

John Zachary Young 1907–

9   In the academic warfare between arts and sciences the linguists are the shock troops of the arts faculty. Only they have an arcane terminology and grasp of statistics adequate to repel the invasion of scientists into their territory. Unfortunately it has kept the scientists at bay, pinned down by such labels as 'Behaviourists' or 'Reductionists'.
    Programs of the Brain 1978 (Oxford: Oxford University Press)

10   What would be the use of a neuroscience that cannot [could not!] tell us anything about love?
Programs of the Brain 1978 (Oxford: Oxford University Press)

Thomas Young 1773–1829

11   We must recollect that every analysis of an unknown object of this nature must unavoidably proceed more or less by the imperfect argumentation sometimes very properly called a circle, but which, in such instances, may be more aptly compared to a spiral, or to an algebraical approximation; since, by assuming certain incorrect suppositions, not too remote from the truth, we may render them, by means of continual repetition of the calculation, more and more accurate, until at length the error is rendered wholly inconsiderable.
Encylopaedia Britannica, entry EGYPT in supplement to the 4th, 5th and 6th editions. 1816–1824

12   Much as I venerate the name of Newton, I am not obliged to believe that he was infallible. I see...with regret that he was liable to err, and that his authority has, perhaps, sometimes even retarded the progress of science.
1801

Yuan Mei 1716–1797

13   When the lists go up much is heard of the candidates' resentment; No one realizes with what sadness the examiners did their duty.
Arthur Waley *Yuan Mei* 1956 (London: Allen & Unwin)

Leonid Mitiofanovich Zamyatin 1922–

1   [USSR Ambassador to Britain]
    I do not think it would be an exaggeration to say that it was perhaps a
    small personal computer that triggered a lot of important changes in the
    minds of many politicians.
    The Guardian 25 April 1990

Yevgeni Ivanovich Zamyatin 1884–1937

2   Tell me, what is the final integer, the one at the very top, the biggest of
    all?
    But that's ridiculous! Since the number of integers is infinite, how can you
    have a final integer?
    Well then how can you have a final revolution?
    There is no final revolution. Revolutions are infinite.
    We transl W N Vickery in P Blake and M Hayward *Dissonant Voices in Soviet Lit-
    erature* 1962 (New York: Random House)

Alfred Zauberman

3   [London School of Economics]
    Zauberman's Law:
    The worse the economy, the better the economists.
    The Guardian 5 October 1983

E Christopher Zeeman 1925–

4   Technical skill is mastery of complexity while creativity is mastery of sim-
    plicity.
    Catastrophe Theory, Selected Papers 1972–77 1977 (Reading, MA: Addison-Wesley)

John Ziman 1925–

5   ...the specialised originality of the mind of the research scholar interacts
    with the unsophisticated radicalism of the student, generating and sustain-
    ing the tradition of a love of learning, intellectual integrity, and boldness
    of thought.
    Science 22 September 1978 **201** 1115

6   Ideas move around inside people.
    [Through travel by Boeing 707, rather than as mental borborygm]
    The Force of Knowledge 1976 (Cambridge: Cambridge University Press)

7   Science is public knowledge.
    Public Knowledge; the Social Dimension of Knowledge 1968 (Cambridge: Cambridge
    University Press)

Solly [Lord] Zuckerman 1904–

8   The nuclear world, with all its perils, is the scientists' creation; it is certainly
    not a world that came about in response to any external demand.
    Apocalypse Now? 1980 (Nottingham)

Index

Acknowledgments

Grateful acknowledgment is made to the following for their kind permission to reprint copyright material. Every effort has been made to trace copyright ownership but if, inadvertently, any mistake or omission has occurred, full apologies are herewith tendered.

Full references to authors, the titles of their works, and publishers are given under the appropriate quotation.

Addison–Wesley Publishing Co, London

Akademische Verlagsgesellschaft, Leipzig

George Allen & Unwin Ltd, Hemel Hempstead

American Association of Physics Teachers, New York

American Institute of Physics, New York

American Physical Society, New York

Architectural Press Ltd, London

Associated Book Publishers Ltd, London

Association of University Teachers, London

Basic Books Inc, New York

Baskervilles Investments Ltd, London

Professor Stafford Beer

G Bell & Sons Ltd, London

Adam & Charles Black Ltd, London

The Bodley Head, London

British Association for the Advancement of Science, London

British Broadcasting Corporation (Publications), London

Rita Bronowski for the Estate of Jacob Bronowski

The Bulletin of the Atomic Scientists, Chicago, Illinois

John Calder (Publishers) Ltd, London

Cambridge University Press, London

Campbell Thomson & McLaughlin Ltd, London

Jonathan Cape Ltd, London

Cassell & Collier Macmillan Publishers Ltd, London

Chatto & Windus Ltd, London

The Ciba Foundation, London

Ronald W Clark

Arthur C Clarke

Miss D E Collins

William Collins, Sons & Co Ltd, London

Committee of Vice-Chancellors and Principals of the Universities of the United Kingdom, London

Constable & Co Ltd, London

Curtis Brown Ltd, London

J M Dent & Sons Ltd, London

Andre Deutsch Ltd, London

Diogenes Paris

Elizabeth H Dos Passos

Doubleday & Company Inc, New York

Dover Publications Inc, New York

The Economist Newspaper Limited, London

Edinburgh University Press, Edinburgh

Edition Stock, Paris

Editions Gallimard, Paris

Elek Books Ltd, London

Elsevier Publishing Co, Barking

Encounter Encounter Ltd, London

Encyclopaedia Britannica Inc, Chicago, Illinois

D J Enright

Estate of Mrs George Bambridge

Estate of E C Bentley

Estate of Albert Einstein

Estate of Robert Frost

Estate of C Day Lewis

Estate of George Orwell

Estate of Bertrand Russell

Estate of Lord Tweedsmuir

Estate of H G Wells

Evening Standard Beaverbrook Newspapers Ltd, London

Faber & Faber Ltd, London

R Buckminster Fuller

Gaberbocchus Press Ltd, London

Victor Gollancz Ltd, London

Professor Samuel A Goudsmit

Robert Graves

Warren H Green Inc, St Louis, Missouri

The Guardian London

Mrs Helen Spurway Haldane

Hamish Hamilton Ltd, London

The Hamlyn Publishing Group Limited, Feltham

Harcourt Brace Jovanovich, Inc, New York

Harper & Row, Publishers, Inc, New York

Thomas G Hart

Harvard University Press, Cambridge, Massachusetts

A M Heath & Co Ltd, London

William Heinemann Ltd, London

Heinemann Educational Books Ltd, London

Her Majesty's Stationery Office, London

David Higham Associates Ltd, London

History of Science Society, Washington, DC

The Hogarth Press Ltd, London

Holt, Rinehart and Winston Inc, New York

Houghton Mifflin Co, Boston, Massachusetts

Humanities Press Inc, Atlantic Highlands, NJ

Hutchinson Publishing Group Ltd, London

Mrs Laura Huxley

IEEE Spectrum New York

Institute of Theoretical Physics, Calcutta

International Statistical Institute, Voorburg

The International Union of Biological Sciences, Paris

The Johns Hopkins University Press, Baltimore, Maryland

Mrs Katherine Jones

Professor R V Jones

Alfred A Knopf Inc (a division of Random House Inc), New York

The Lancet London

Lawrence & Wishart Ltd, London

Tom Lehrer

Claude Levi-Strauss

Librairie Hachette, Paris

Librairie Plon, Paris

Arthur D Little Inc, Cambridge, Massachusetts

Little, Brown & Co, Boston, Massachusetts

Macmillan & Co Ltd, London and Basingstoke

Macmillan Publishing Co Inc, New York

Sir Peter Medawar

The Medical Journal of Australia Glebe, NSW

The Merlin Press Ltd, London

Julian Messner Inc (a division of Simon & Schuster Inc), New York

Minerva London

The MIT Press, Cambridge, Massachusetts

William Morrow & Co Inc, New York

John Murray (Publishers) Ltd, London

National Aeronautics and Space Administration, Washington, DC

Nature Macmillan (Journals) Ltd, London

Dr J Needham

The New American Library Inc, New York

New Directions Publishing Co, New York

New Scientist (the weekly review of science and technology) New Science Publications, London

The New York Academy of Sciences, New York

The New York Review of Books Copyright ©1970 Nyrev, Inc, New York

New York University Press, New York

The New Yorker New York

North-Holland Publishing Company, Amsterdam

W W Norton & Company Inc, New York

Harold Ober Associates Inc, New York

The Observer London

Orell Füssli Verlag, Zurich

Mrs Sonia Brownell Orwell

Peter Owen Ltd, London

Oxford University Press, New York

Oxford University Press, Oxford

Pantheon Books Inc (a division of Random House Inc), New York

C Northcote Parkinson

Pemberton Publishing Co Ltd, London

Penguin Books Ltd, London and Harmondsworth

Pergamon Press Ltd, Oxford

A D Peters & Co, London

Phaidon Press Ltd, London

The Physical Review Long Island, NY

Physical Review Letters Long Island, NY

Professor Sir Brian Pippard

Pitman Publishing Ltd, London

Praeger Publishers Inc, New York

Presses Universitaires de France, Paris

Princeton University Press, Princeton, NJ

G P Putnam's Sons, New York

Random House Inc, New York

D Reidel Publishing Company, Dordrecht

Revista de Occidente SA, Madrid

Kenneth Rose

Routledge & Kegan Paul Ltd, Henley-on-Thames

Royal Geographical Society, London

The Royal Institution, London

The Royal Society, London

Salk Institute, San Diego, California

Sather Classical Lectures ©1952 by the Regents of the University of California

Saturday Review New York

John Schaffner Literary Agency, New York

Schenkman Publishing Company, Cambridge, Massachusetts

Science American Association for the Advancement of Science, Washington DC

Science Policy Foundation, London

Scientific American W H Freeman & Co, San Francisco, California

SCM Press Ltd, London

The Scotsman Edinburgh

Charles Scribner's Sons, New York

Martin Secker & Warburg Ltd, London

Sheed & Ward Ltd, London

Simon & Schuster Inc, New York

Martyn Skinner

The Society of Authors for the Estates of Gordon Bottomley, John Masefield and Bernard Shaw

The Society of the Friends of Chester Cathedral

Souvenir Press Ltd, London

The Spectator London

Spiegel-Verlag, Hamburg

Springer-Verlag, New York

Stanford University Press, Stanford, California

Sunday Telegraph London

Stefan Themerson

Charles C Thomas, Publisher, Springfield, Illinois

H W Tilman

Time Magazine The Weekly News Magazine ©Time Inc 1976

The Times Times Newspapers Ltd, London

UNESCO, Paris

University of California Press, Berkeley, California

The University of Chicago Press, Chicago, Illinois

University of London Bulletin London

University of Pennsylvania Press, Philadelphia, Pennsylvania

University of Toronto Press, Toronto

Mrs M J Waddington for the Estate of Professor C H Waddington

Walker & Co, New York

A P Watt & Son, London

C A Watts & Co Ltd, London

Weidenfeld & Nicolson Ltd, London

John Wiley & Sons Inc, New York

John Wiley & Sons Ltd, Chichester

World Federation of Scientific Workers, London

Miss Anne Yeats

M B Yeats

Znanie, Moscow